D1013218

THE LITTLE BOOK
of stars

james b. kaler

C

COPERNICUS BOOKS

AN IMPRINT OF SPRINGER-VERLAG

Published in the United States by Copernicus Books,
an imprint of Springer-Verlag New York, Inc.
A member of BertelsmannSpringer Science+Business Media GmbH

Copernicus Books
37 East 7th Street
New York, NY 10003
www.copernicusbooks.com

Library of Congress Cataloging-in-Publication Data
Kaler, James B.
     The little book of stars / James B. Kaler.
        p.   cm.
     Includes index.
     ISBN 0-387-95005-2 (alk. paper)
     1. Stars.   I. Title.
     QB801.K24   2000
     523.8—dc21   00-031848

Manufactured in the United States of America.
Printed on acid-free paper.

9 8 7 6 5 4 3 2 1

ISBN 0-387-95005-2   SPIN 10755958

to my cousins
URSULA SCHUSTER and DAVID KALER
with thanks for a lifetime of love and friendship

acknowledgments

Many are the stars of this book. Thanks go first to Jerry Lyons for the ideas and support that led to the concept and to its early development; to Editor-in-Chief Paul Farrell, who provided wonderful additional support and shepherded the book through to its completion; and to Publisher Kevin Lippert, again for support and for ensuring a quality product. Two readers provided essential feedback for revision of the first draft. Karen Kwitter of Williams College reviewed the manuscript for scientific accuracy and writing quality, while Ursula Schuster played the role of interested reader and kept me on the path of clarity. Janice Borzendowski did a masterful job of editing, Jordan Rosenblum of designing, and Keithley and Associates of developing the art, while Editorial Assistant Mareike Paessler kept things moving at a brisk and efficient pace. Thanks to all, and to my wife Maxine for her continued support.

preface

The Little Book of Stars tells the story of stellar science and what the stars mean to us from a variety of perspectives. Beginning with the "big picture," the book moves through progressively more and more intimate views until we feel we can hold the stars in our hands, from which we can then throw them back to the sky to see our place among them.

The book opens with a summary of the event that created our Universe, the Big Bang, and then goes on to describe the natures of the Big Bang's progeny, the stars—what they are, how they shine, and how they can live such immensely long lives. Approaching home, it next examines the measures of the stars: where they are, how they are collected together from pairs to galaxies of billions, and how we learn of their individual properties. Yet closer, we look in depth at the Sun and at the physical differences among the stars, at the immense range of properties they possess. Finally, arriving at Earth, we see the significance of the stars to human life, how we have used them to tell our stories and to find where we are in both space and time.

From this base, the book looks more closely at stellar details, concentrating on temporal phenomena—on stellar change—and on the observational base that helps set the stage for the theory that links them all together. Here, in the theory of stellar evolution, we see the fate of the Sun and of all the other stars, and how the great array of stellar properties is created. The end returns us to the beginning, to the births of stars from the cold depths of space and from the deaths of previous stellar generations. We now see ourselves reflected in starlight as we watch the formation of our own Sun and planet; here we see that we are tightly linked to the Universe at large and that we are a direct part of its family. Enjoy the tour and then, above all, go outside and watch them glitter for you in the nighttime sky.

— *J.B.K.*
Urbana, Illinois
September 2000

contents

stars

Day and night. Our lives are run by the flow of one to the other, the distinction as profound as anything we know. Daylight, so bright we shield our eyes, not daring to look at the brilliant source of illumination, the Sun, perceived through the ages as godlike. Night, though dark, not black at all, the sky awash with thousands of shimmering lights. Day and night: both reflect the inescapable stars, those at night far away, the one that lights the day nearby, the heat of the Sun telling of the immense concentrations of energy that are the stars. What are they? Where are they? How do they shine? How did they come to be? What might happen to them? What do they mean to us and how do they relate to our life-giving Sun?

The story of how we learned the stars' natures belongs to our own times, written from ever-accelerating study and understanding during the past 400 years. The story of the stars themselves, of their actual natures, goes back vastly farther, to the beginnings of time itself, to the event that created the Universe, to the Big Bang.

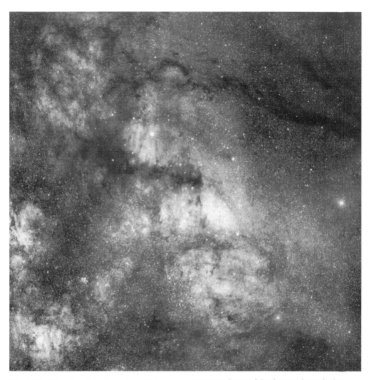

The Milky Way, the disk of our huge Galaxy, sends showers of stars through the constellations of Ophiuchus and Scorpius. Each star has a different story to tell. Some are huge, some small, some young, some old. Each plays a role in begetting the next stellar generation, our own Sun and Earth a product of what has gone on before. (*Atlas of the Milky Way*, F. E. Ross and M. R. Calvert, University of Chicago Press, 1934. Copyright Part I by the University of Chicago. All rights reserved. Published June 1934.)

in the beginning…

Stars are the main repository of illuminating—bright—matter in the Universe. From them comes most of our light. Like raindrops from a thundercloud, they are the condensates of the Big Bang. The "Big Bang," though a scientific misnomer, provides a wonderful metaphor that gives some sense of what happened so long ago. Look outward past the night's stars, which in the grand scheme are all "local," all belonging to our Galaxy of 200 billion of them. In the depths of space, we find countless more galaxies, trillions of them, assemblies in which billions of stars can be caught together in a single glance.

Distances are incomprehensible, measured by the light year, the distance a beam of light travels in a year at a speed of 300,000 kilometers (186,000 miles) per second. The nearest star is 4 light years away, seen as it was 4 years ago. Our Galaxy—a disk-shaped structure that manifests itself at night as the Milky Way—is 80,000 light years across; entire ice ages can come and go while light travels from one side of it to the other. The nearest large galaxy lies at 25 times that distance; vast numbers of galaxies extend to billions of light years away. All are receding from us at speeds that increase steadily with distance, as if blasted outward from some great explosion, the fastest-moving pieces now the farthest. We appear to be at the center of it all, until we realize that anyone in any galaxy out there would see the same thing—all systems moving away from all other systems unless they are close enough for gravity to keep them together.

Go back in time to when they—or what they were to be made from—were all together. From the measured expansion rate of 20 kilometers (12 miles) per second per million light years of distance, the expansive event from which they were flung happened somewhat less than 14 billion years ago. At that moment, at the beginning of time, there could not have been separate stars or galaxies. The entire

Two thousand galaxies within the "Hubble Deep Field" (less across than a tenth the angular size of the Moon, only a portion seen here) stretch away to unimaginably great distances, to billions of light years. Each galaxy approaches a hundred thousand light years in size and is home to billions of stars. Each is moving away from us—and from every other—as a result of the Big Bang, the event that created what we consider our Universe. We are a product not just of our own Galaxy but of the entire Universe. (R. Williams, the HDF Team, StScI, and NASA.)

Universe consisted not of matter as we know it, but of energy. Mass and energy are dual entities that can be converted back and forth into each other. Where the Universe and its contents came from, no one knows. Perhaps it all sprang whole from the eternal vacuum. The temperature at the beginning—or as close to the beginning as we can get, within the tiniest fraction of a second—was a billion trillion trillion degrees Celsius.

As a result of its vast energy, the Universe could do nothing but swell with great speed and, as in any expanding system, cool. As the temperature dropped, within only a fraction of a second after the beginning, much of the energy froze into what we know as matter, which thereafter continued to be hurled outward. The event was not an explosion that expelled mass and energy through space; it was— and still is—an expansion of *space itself*, in which matter is caught to float like clouds moving with a wind, an expansive wind that moves them steadily apart. As the temperature chilled further, still before a second of time had elapsed, matter condensed from its first primitive state into now-familiar atomic constituents—protons, neutrons, and electrons.

Protons are particles that carry positive electric charges. They have diameters a mere tenth of a trillionth (10^{-13}) of a centimeter (a millionth would be 10^{-6}, a trillionth 10^{-12}) and weigh in at only a trillionth of a trillionth (10^{-24}) of a gram. Neutrons are similar in size and mass but carry no charge. Electrons are much smaller and have a charge equal to that of the proton, but negative instead of positive. The chemical elements are constructed of these particles, consisting of atoms with positive, proton-neutron nuclei surrounded by negative electrons. The kind of element depends on the number of protons in the nucleus; hydrogen has 1, helium 2, oxygen 8, iron 26, and so on. At low temperatures, each positive proton in an atom is balanced by a negative electron (so that you suffer no lethal shock when you touch something).

Protons, by themselves, constitute the simplest of all atomic nuclei, hydrogen. At an age of some three minutes, the still high, though falling, temperature and density of the Universe conspired to slam hydrogen's protons and neutrons into each other so as to freeze some of them into most of the Universe's helium (whose normal nucleus consists of two protons and two neutrons each), and a bit of lithium, providing the raw material—92 percent hydrogen and 8 percent helium—out of which the stars would someday form.

The temperature at this time was still so high that the young Universe remained a sea of free atomic nuclei and electrons, which allowed energy and mass to interact fiercely with each other. But 100,000 years later, when the expansion had driven the temperature down to 7000 degrees Kelvin (Celsius degrees above absolute zero)[*], the electrons and nuclei combined, enabling the remaining energy, in the form of light, to roam free.

Light is a generic term for radiation that carries energy in the form of electromagnetic waves (sometimes thought of as wave-particles called *photons*) from one place in the Universe to another. The amount of energy that light carries depends directly on the *frequency* of the waves (the number passing per second), or inversely on their length (the distance between the wave peaks), the *wavelength.* The shorter the waves, the greater the energy. The wavelengths of visible light, which determine the colors we see, are quite small, ranging from around 0.00004 centimeter for extreme violet light to 0.00008 centimeter for extreme red. With its shorter waves, violet light packs twice the energy of red. The shortest waves, those with the highest energies, thousands of times that of visible light, are gamma rays; in

[*] Absolute zero is –273 degrees Celsius (–459 degrees Fahrenheit). The Kelvin scale, named after the British scientist William Thomson (Lord Kelvin), counts Celsius (Centigrade) degrees upward from absolute zero. Water freezes at 273 degrees Kelvin and boils at 373 degrees Kelvin.

between are familiar body-penetrating X-rays, and just to the high-energy side of violet is the ultraviolet. (These are all quite dangerous; even ultraviolet causes burns or worse—sunburn comes from a small bit of solar ultraviolet that gets through our atmosphere, which is opaque to most high-energy radiation). Off the red end of visible light is the infrared; waves thousands of times longer than red are termed *radio*.

Temperature reflects the energy inherent in matter. High-temperature matter produces all kinds of radiation, including high-energy gamma rays and X-rays. At low temperatures, only low-energy radiation, radio, can be produced. The Sun, at 5780 degrees Kelvin, copiously radiates in the middle, at "visual" wavelengths to which the eye is sensitive. Today, we see the Big Bang's radiation all around us as low-energy radio waves coming from a Universe cooled to only 3 degrees Kelvin. The Big Bang, in which the Universe erupted from a hot dense state, was postulated from the observed current expansion. Over 50 years ago, astronomers predicted it should have cooled to near its current temperature. The discovery of this chilled radiation in 1965 magnificently supported the theory. More than any other evidence, the cosmic background radiation tells us that the Big Bang really happened.

Though the Universe is dominated by its expansion, the driving force behind the creation of stars and galaxies is gravity. Random fluctuations in the early matter of the expanding Universe could, through the attraction of gravity, grow to larger units. Just what these first units were we are not sure. They could have been relatively small blobs of matter that created stars within themselves, or perhaps stars that accumulated into primitive small galaxies. Whatever they were, the first growth took place quickly, within a billion years after the Big Bang. Smaller galaxies then grew larger through collisions and mergers to make those we see around us today.

When we look at any object in space, we see it not as it is, but as it was: your finger a trillionth of a second ago, Alpha Centauri four years ago, nearby galaxies millions of years ago. For nearby astronomical bodies, light-travel time is not important, as stars typically live, and galaxies evolve, over billions of years. But as we look outward to very distant galaxies, we look back close to the creation event itself and can see what things were like in the Universe's early days. With great telescopes we can actually see small distant galaxies making early generations of stars, and can see mergers as they were taking place.

Yet so much remains to be learned. Much of the matter that it took to create the stars and galaxies is still in unrecognized and mysterious form. Called *dark matter,* it is known to us only through its gravity. Only 1 percent or so of the mass of the Universe seems to be made of stars, but that small statistic belies the stars' importance. Whatever the dark matter may be, the brilliant stars trace it, show us where it is. Stars also drive the evolution of galaxies, and in that sense part of the evolution of the Universe itself. Stars do not exist in isolation, but affect each other. Indeed, the natures and births of stars are directly dependent on the deaths of others. Through a variety of interactions, the lives of the stars led to the birth of one in particular, the one that brings daylight, the one that gave us birth, the one that made the study of the stars and of the Universe possible in the first place.

stars defined

Everyone knows of stars. They are seen even from brightly lit cities. It would seem easy to define them as "lights in the sky that are not planets or other bodies of our Solar System." But there are so many different kinds of stars, so many that cannot be seen with the naked eye, some so bizarre as to strain belief, that an accurate definition requires an introduction to some of their properties, and especially to what happens inside to make them shine.

In the simplest terms, stars are self-luminous balls of hot gas, whereas planets and the like glow mostly with light they reflect from the Sun. Most, but not all, stars are large. The Sun, a rather typical star, dwarfs Earth. It is 1.5 million kilometers, 100 Earths, across, and contains the volume of a *million* Earths. None of the other planets can compete either; the Sun is 10 times the diameter of the biggest planet, Jupiter. Even at a distance of 150 million kilometers (the *Astronomical Unit*, or AU, equal to just over 100 solar diameters), the angular size of the Sun is half a degree across in our sky (but do not dare look at it).

Yet as big as they are, the stars are small compared with the spaces between them. That the sky seems filled with stars is an illusion. The nearest nighttime star, Alpha Centauri, is almost 300,000 times farther from Earth than is the Sun. Light comes to us from the Sun in 8 minutes; from Alpha Centauri it takes 4 years. Where we live, in the outer part of the Galaxy, stars are separated by more than 10 million times their diameters. We need not worry about collisions.

Stars are hot. The surface of the Sun shines at nearly 6000 degrees Kelvin, 20 times the temperature of Earth. Some stars are 100 times hotter. Any body within a cooler surrounding will attempt to get rid of its internal energy by radiating energy as various forms of light (from radio to gamma-ray). Hot bodies therefore glow visibly on their own. For a given size, the greater their temperature, the greater their brightness. Experiment and theory agree that the brightness of a stellar surface depends on the fourth power of its temperature: double the temperature and the star becomes 16 (or 2^4) times brighter; triple it and it radiates 81 (or 3^4) times as much. High temperature coupled with great size makes for great luminosity. The Sun radiates the astonishing power of 400 million million million million (400×10^{24}, where a million is 10^6 or 1,000,000) standard 100-watt light bulbs. The number is so large, the energy so great, that astronomers use the Sun itself as the standard unit against which to measure all other stars. How else could you feel not just warmth, but intense heat, from a body 150 million kilometers away? The range in luminosity among stars is equally awesome. Some stars are a million times brighter than the Sun, while others are a million times dimmer. Masses are huge, the Sun containing 330,000 times more matter than Earth and 1000 times that of Jupiter, indeed almost 1000 times the mass of the entire planetary system.

Now to the heart of it: age. The stars seem ageless. The night sky exhibits almost no change over a human lifetime. Similarly, aside from what we now know were exploding stars, the entire written human record reveals no evidence of long-term change in distant stars or in the Sun. And Earth's fossil record, which takes us back hundreds of millions of years, tells us the Sun has been shining about the same as now for all that time.

To explore farther back we use the atomic record. Most atoms are stable; hydrogen and helium have been here since the Big Bang created them. But other atoms are not so lucky. We designate a chemical element—whether iron, oxygen, or gold—according to the number of protons in its nucleus. The number of protons in any element is thus set—defining the element. The number of attached neutrons for each element, in contrast, can differ, producing a variety

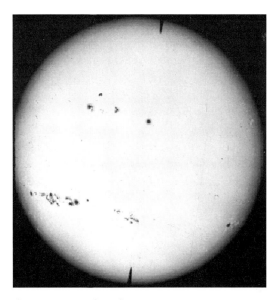

The Sun, our own (and typical) star, dwarfs us. The larger sunspots, regions of intense magnetism that temporarily chill local areas of the solar surface, are bigger than Earth. We are a by-product of the Sun's birth and are totally dependent on it for life. (The Observatories of the Carnegie Institution of Washington.)

of *isotopes*. The simplest isotope of hydrogen consists of a single proton, but the proton can also couple to a neutron to make an isotopic form called *deuterium*, or because it has two bound particles, hydrogen-2. (Deuterium was also created in the Big Bang. In about 0.001 percent of the water you drink, a normal hydrogen atom is replaced by deuterium: natural hydrogen-deuterium-oxide.) Now, however, add a second neutron to make hydrogen into tritium (hydrogen-3), and the atom *decays*: that is, it becomes something else. (A neutron in the tritium nucleus turns into a proton by ejecting a negative electron, which makes stable helium-3; most helium, by far, is helium-4.) Most elements have a small range of allowed neutrons, and thus a small range of stable isotopes. With elements heavier than bismuth (83 protons), however, *no* stability is allowed, and all atoms decay into lighter stable ones.

As unstable isotopes disintegrate, they release radiation in the form of gamma rays and atomic particles; that is, they are *radioactive*, some dangerously so. Uranium, radium, cobalt-60, and strontium-90 are deadly examples. In nature, radioactive decay does not take place all at once. (When uranium decay is induced to take place all at once, the result is an atomic bomb.) Any radioactive isotope has a natural lifetime over which it decays; a kilogram of uranium-238 (92 protons and 146 neutrons, adding to 238 particles) will decay to half a kilogram in 4.5 billion years, passing through radium and down to an isotope of lead. Rock contains a variety of radioactive elements. By comparing the amount of a final decay product with that of the initial radioactive isotope, the daughter with the parent, and by knowing the isotope's lifetime from lab measurements, we can date a rock since solidification (when the elements were sealed within it). The oldest Earth rocks come in at 3.5 billion years, Moon rocks and meteorites (small asteroids that hit the Earth) at 4.5 billion. But what does this tell us about the Sun? Watch the nighttime sky and you see the planets revolve around the Sun against the distant background of the stars. They all move through the ancient constellations of the zodiac, which lie along the *ecliptic*, the path of the orbiting Earth as viewed from the Sun, which consequently is also the apparent path of the Sun around the moving Earth. Planetary orbits are thus closely confined to a plane centered on the Sun. Moreover, all orbit the Sun in the same direction, that of solar rotation, and nearly in the Sun's rotational plane. It became evident long ago that the planets must have formed from a spinning gaseous disk centered on the forming Sun, meaning that all should be about the same age. The Sun therefore must be about as old as the oldest members of the Solar System, or 4.5 billion years. (That the Earth seems younger than some other bodies of the Solar System just means that it took a longer time for it to cool and solidify after formation.) By analogy, other stars must be of a somewhat comparable age.

It is easy to explain how a ball of gas as big as the Sun—or any other star—radiates as much as it does. Compress a gas and it heats; heat a gas and it wants to expand (driving pistons and moving automobiles down the road). In a car, compressive force is supplied by the flywheel, in the Sun, by gravity. The compression drives up the temperature until the outward expansive pressure exactly balances the inward pull. The compression raises the internal pressure and temperature—like the mixture in an engine cylinder—to immense amounts. Given the known laws that govern the behavior of gases, astronomers calculate that at the center of the Sun the temperature is 15 million degrees Kelvin and the density is 13 times that of lead. Yet the solar center is still gaseous! These are the conditions that are needed to produce the observed solar luminosity.

But we cannot get something for nothing (perpetual motion machines do not exist). The radiated energy must come from somewhere. Drop a bowling ball and you release energy, compliments of gravity (if you doubt that, drop it on your foot). The Sun's energy could come from gravity alone, but as the Sun radiated, it would have to "fall," in the sense of shrink. Radiation acts to cool the Sun; slow shrinkage would then maintain the internal temperature. Gravity alone can therefore produce the Sun's luminosity. But given the total gravitational energy available, it could do so for only 10 million years, far short of the 5 billion years we have already determined as the Sun's age. Something else must be at work.

Though there are great numbers of different kinds of stars, analysis of starlight has shown that the great majority, including the Sun, are made almost entirely of the simplest of atoms, those of hydrogen and helium supplied by the Big Bang. Only 1 atom out of 1000 is something heavier, and within this tiny minority oxygen and carbon dominate. Helium is made of four particles, in its most common state of two protons and two neutrons. A helium atom, however, weighs a bit less than the four particles do alone, the whole rather remarkably less than the sum of its parts. If we can somehow combine two protons and two neutrons to make helium, we lose mass. In 1914, Albert Einstein discovered that mass and energy are equivalent, the two convertible via the famed formula: $E = Mc^2$. If we multiply the amount of mass (M) by the square of the speed of light (c, a very large number), we obtain a great amount of energy (E). In the moments of the early Big Bang, energy converted to matter. Now we see the opposite, matter converting back to energy. The mass lost in creating helium goes into the energy that is radiated by the star. But solar neutrons and protons cannot simply combine to make helium as they did in the early Big Bang because free neutrons—those not already in an atom—decay quickly and are a rarity in a stellar gas. Our raw material must be protons alone, found in abundance as simple hydrogen. Four atoms of hydrogen must somehow conspire to make one of helium, with 0.7 percent of the mass lost in the process.

Gravity is therefore not directly needed for energy production, so the Sun and other stars—as long as they are making helium—do not shrink as they would if gravity acted alone. Fusion of hydrogen to helium can work only under high temperatures and densities, however. As a result, the process is called *thermonuclear fusion*. Gravity's job is to supply the high temperature in the first place, the temperature that makes the fusion possible. The solar mass and luminosity coupled to Einstein's equation shows that there was enough hot interior hydrogen to run the Sun, and to keep gravity at bay, for 10 billion years. Since the Sun is almost 5 billion years old, it has run through about half its lifetime. The shock here is that stars do not last forever. They die, they all die, some quietly, some with unimaginable violence. The good news is that, at least for the Sun, the event is still 5 billion years off. The bad news is that you still have to go to work.

Now, at last, we can define a star. It is a body that holds itself together by gravity and that: (1) is now running on thermonuclear fusion; (2) once ran on thermonuclear fusion (which allows us to call dead stars "stars"); or (3) will someday run on thermonuclear fusion (allowing us to call stars in formation "stars"). The definition is simplicity itself: thermonuclear fusion is all.

how fusion works

There are four known forces of nature that act over a distance, and all have solar roles. Gravity, the first and most familiar, is surprisingly the weakest. All matter produces this always-attractive force; all matter attracts all other matter, trying to draw itself together. The Universe exists *because* it is expanding. Were it not, gravity would make it contract, and we would be crushed out of existence.

Theoreticians recognized this concept before Edwin Hubble discovered the expansion in 1929. Yet atom to atom, proton to proton, gravity is exceedingly weak. We stay tied to Earth, Earth orbits the Sun, and stars are kept together only because there are so *many* atoms involved: 10^{28} in ourselves, 10^{50} in the Earth, 10^{57} in the Sun, every one attracting every other one over any and all distances.

Electricity and magnetism combine to make the electromagnetic force, the second force, which is vastly more powerful than gravity. We generally pay it little heed unless it becomes unbalanced by removing electrons from atoms, which charges a body and allows a flow of electricity. The electromagnetic force envelopes us, however; it makes possible the chemical reactions that produce the countless kinds of molecules (combinations of atoms) of which our world is made, and gives light and substance to our world. Atoms and molecules are almost all empty space. The proton of hydrogen is a grain of rice at the center of a sports stadium, cheered on by the electron in the outer seats. What we feel as solid matter is the electromagnetic force, which pervades the atom and keeps the electrons tied to the protons. It is this force that produces radiation, electromagnetic radiation, of which visible light is just a small part.

The other two forces are more obscure. Protons and neutrons at first appear similar, almost the same weight, the only difference appearing to be charge. Protons left alone are forever. The protons within the hydrogen that makes up so much of our bodies (in water, for example) were for the most part created within the first second of the Big Bang, of which we are a true product. Neutrons, in contrast, can be ephemeral. Leave a bare neutron alone and in 15 minutes there is a 50:50 chance it will be gone, having spat out a negative electron to become positive—that is, to become a proton. Sealed in an atom the neutron can be permanent, but in a radioactive atom it (as well as the proton) can still transform itself. These transitions are a product of the *weak force* (weaker than electromagnetism), with which we come into contact in radioactivity.

Similar charges attract, opposites repel, the reason the electrons hold to their atoms. So how can two protons bind together in a helium atom? Enter the fourth force, the *strong force*. Unlike the electromagnetic and gravitational forces, which act over (though weakening with) all distance, this force acts over only a short range: the size of the particle, either proton or neutron (electrons do not carry the strong force). But within that limit it is supreme. Gather protons and neutrons into a tight embrace, and even the power of electromagnetic repulsion cannot drive them apart.

The strong force provides the key to understanding thermonuclear fusion. Dive deep into the core of the Sun, where the temperature is above 10 million degrees Kelvin. The speeds of atoms respond to temperature. In your room, they are moving at a kilometer per second, the reason you feel warm. In the Sun, protons typically move 200 times faster, and a very few skim along vastly faster than that. An ultrafast proton can slam into another so hard that electromagnetic repulsion cannot keep the pair apart, and they approach close enough for the strong force to try to make them stick. But even the strong force cannot keep a pair of mutually repulsive protons together. Now enter the weak force. As the neutron can release an electron to make a proton, so one of the bound protons can kick out a *positive* electron to become a neutron. The neutron is then glued by the strong force to the proton. The center of the Sun is making deuterium from hydrogen.

The positive electron—a "positron"—is antimatter! No science fiction here: antimatter, matter with reversed properties (especially charge), exists. Fortunately it is not found in the Universe in any great abundance, because matter in contact with antimatter annihilates to become electromagnetic energy. (The classic test for an antimatter planet would be to land an astronaut: if he or she blows up . . .) The positron goes hardly anywhere before meeting a free electron, at which point the two convert to a pair of high-energy gamma rays. The binding of the protons into stable deuterium thus creates electromagnetic energy. The heat absorbed from the gamma rays slowly works its way out by radiation and finally appears at the solar surface as a yellow-white light. (Direct gamma rays would kill everything. Think of the sunburn caused by ultraviolet light, which is only a bit more energetic than visible violet.)

As yet we have no helium. Almost as soon as the deuterium is formed, it will pick up another proton to make light helium (rare helium-3, also made in the Big Bang). Finally, two helium-3 atoms collide and partly shatter. The debris consists of a helium-4 atom (the common kind of helium) and a pair of protons, which scuttle away, later to try fusion all over again. The result is that four atoms of hydrogen are converted to one atom of helium, and mass is lost and energy made in the binding. Since hydrogen atoms are protons, the total reaction set is called the *proton-proton chain*.

If all this seems far-fetched, we have direct proof. In the "laboratory" (used in the larger sense), we can make heavy forms of hydrogen combine together very quickly to make helium. The result is a hydrogen bomb. More to the point, the conversion of the proton into a neutron and a positron also makes an energy-carrying particle called the *neutrino*. It takes tens of thousands of years for electromagnetic energy to work its way out of the Sun, because it is successively re-absorbed and re-emitted by atoms. But the neutrinos, because they are weak-force particles, do not much sense the surrounding gas and escape the solar center directly and immediately, moving at nearly the speed of light. (Billions of solar neutrinos pass through you each second with no effect whatever; they even pass through you at night because they sail right through the Earth.)

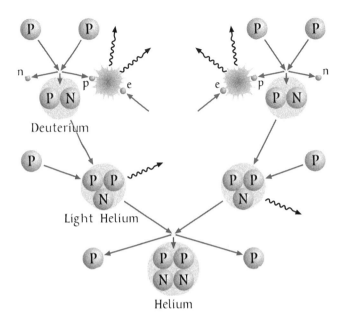

Deuterium

Light Helium

Helium

Like all stars, the Sun shines by thermonuclear fusion, converting hydrogen into helium in its deep core. Helium is lighter than the four hydrogen atoms that make it. The "missing mass" flies out as pure energy. Two fast-moving protons (P) strike and stick, one converting to a neutron (N), to make deuterium. Next, the deuterium picks up another proton to make light helium.

Two light heliums then smash together to make normal helium. High-energy gamma rays (wiggly arrows) are first created when ejected positrons (positive electrons, p) hit normal electrons in mutual annihilation; more are produced by the reaction that makes the light helium. Neutrinos (n) created in the first reaction fly immediately out of the Sun.

Several sophisticated detectors scattered around the world can capture neutrinos, however, making it possible to see them coming from the Sun. Their very presence confirms that the Sun is a nuclear reactor. Given a variety of assumptions, we see them in about the numbers expected from theory. Fusion works.

The proton-proton chain is controlled by the fusion process's first step (proton collision to make deuterium), which fortunately runs very slowly. A typical solar-core proton will live alone for some 5 billion years before it weds another. This aching slowness is a Good Thing, because if the reaction ran fast, the Sun would explode in a grand hydrogen bomb. Nature thereby controls the proton-proton chain, allowing us to enjoy a sunny day.

Other kinds of fusion reactions are possible. At the higher temperatures met in the cores of more massive stars, hydrogen fusion proceeds mostly by the *carbon cycle*. If speeds are high enough, carbon develops a great affinity for protons, absorbing them and making nitrogen. The nitrogen is further converted to oxygen, which then kicks out a helium nucleus and falls back to carbon. Still other reactions that take place in dying and exploding stars fuse helium to carbon and heavy atoms to yet heavier atoms. Stellar winds and explosions carry them away. Everything other than hydrogen and helium (and a taste of lithium) is made in stars. Really.

clash of the titans

Stars exist and function because the forces of nature compete against one another. Stars are born, live, and die as first one force, then another, gets the upper hand. Begin with stellar birth. The space between the stars is far from empty. It is filled with a thin, lumpy, very complex mixture of gas and dust. Yes, dust, solid particles that are too small to be seen with the human eye, made mostly of carbon and silicates combined with other atoms and covered with ices. This dusty *interstellar medium* is thinner than any vacuum created in Earthly laboratories. Yet there is so much space out there that the dust constitutes a significant fraction of the total mass of our Galaxy. Clumped dust clouds block the light of background stars and can be seen with the naked eye as dark structures in the Milky Way.

If a clump of interstellar matter can, through a variety of processes, be sufficiently compressed, its own gravity can make it collapse to form a star. As it contracts, the internal compression heats the interior, and eventually the new star begins to glow. When the core hits 10 million degrees Kelvin, the proton-proton chain begins to produce energy. Nuclear forces now fight against the gravitational force, and eventually prevent the star from shrinking further. Contraction stops and a new star, perhaps one like the Sun, is born. The new star is stable as long as any hydrogen in the core is hot enough to "burn" (in the nuclear sense); the nuclear and gravitational forces are now in stalemate.

When the hydrogen fuel runs out and the core is made of nearly pure helium, energy production through the nuclear (strong and weak) forces dies away. Gravity again gets the upper hand, and the core resumes its contraction. But the clash of forces is not yet over. As hydrogen can fuse to helium, so helium can fuse to carbon, producing yet more energy. Nuclear forces again stop the contraction, but this time not for anywhere near as long. When the helium runs out, the battle tide again turns toward gravity.

The final result depends on the mass of the star. The Sun and most other stars will die when they become small (Earth-sized) balls of carbon and oxygen in which gravity is held off by the strange properties of atomic particles (nuclear fusion now out of the picture). Forces will finally be at peace with one another, and the end

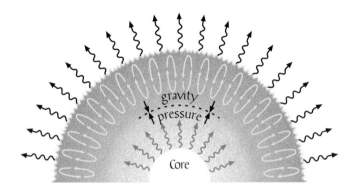

The nuclear reactions that power the Sun and keep it from shrinking under its own gravity take place only in a deep core that houses about 40 percent of the solar mass. From there, the new energy escapes, and is transferred through a thick stable layer that slowly degrades high-energy gamma rays into more benign, longer-wave radiation. Above it is a layer that sends its hot gasses shooting up and down in huge convective overturns, the bubbling cauldron finally releasing its load of light into space at a 6000-degree-Kelvin surface. The Sun everywhere adjusts itself so that the inward pull of gravity is balanced by the outward push of gas pressure, and as a result maintains its size.

product will be stable, essentially forever. In high-mass stars, higher temperatures and densities force gravity and nuclear forces to keep contending. Nuclear burning now runs much further, all the way to iron. Gravity then wins a major battle (though not necessarily the war). The core of a high-mass star, having gone as far as it can under nuclear forces, suddenly collapses. The release of all that stored gravitational energy explodes the star.

The core, however, is so dense that it survives. In one scenario, again depending on mass (or so we think), it will end its life as a stable ball of neutrons about the size of a small city, the forces again at peace. In another, gravity is entirely victorious, and the star collapses forever into the famed and fabled black hole, in which gravity is so strong that light itself cannot escape. Matter blown from stars in the epic battles, either through quiet winds or violent explosions, feeds back into interstellar space, along with freshly made atoms, to fuel yet further star formation in a grand recycling scheme. The Earth and we ourselves are products of it all.

This quick tour of the lives and deaths of stars leaves out the details of battle, the *Sturm und Drang* of daily stellar life. The clash of forces produces some of the most amazing phenomena imaginable, especially when coupled with double stars, whose components orbit each other and interact. In the pages to come we will see stars as big as the Solar System, stars that feed upon each other or even tear each other apart. We will watch stars blowing bubbles and stars that create colorful shrouds worthy of an art gallery. At the same time we will come to understand the stellar peace that allows the existence of planets, including our own Earth, and of life. More, we will see the meaning of stars to ourselves, to humanity. In their parade across the night sky, stars have long linked us to the stories of our ancestors, have long told us how to find our way, and have long told us the time of the day and season. Now they also tell us how we can exist at all.

chapter 2

collections

Stars exist not in isolation but in the company of others. Though not readily apparent to the eye, they form into doubles, triples, and more complex multiples, all held together by gravity. Ascending the hierarchy, singles, doubles, and larger groups merge into clusters, some small and barely discernible, others so obvious as to form constellations. Still others are grander yet, made of millions of stars, these too held together in gravity's grip. Singles, doubles, multiples, and clusters then assemble into our Galaxy, our home, a seemingly simple structure that on closer examination displays remarkable complexity. One of many, our Galaxy is part of a small local cluster of galaxies, which is on the edge of a big cluster of galaxies, which in turn is a part of even bigger structures that help make up the immense and expanding Universe. This too is our home.

location

Whether on Earth or in the sky, the three most important things in real estate are still location, location, location. Before we can really learn much of the stars and their organization, or for that matter of the Universe, we need first to know where the stars and galaxies are, relative both to the Earth and to each other. Though stars and galaxies appear to be plastered on a great dome above our heads, to a mythic *celestial sphere*, like all the world around us they exist in three dimensions. The sky has wonderful depth.

Points in two dimensions—those projected on the sky—are easy to locate, and have been recorded for more than 2000 years. The Earth has an equator set perpendicular to its axis of rotation, the axis emerging at the north and south poles. As the Earth rotates to the east, the sky appears to spin in the opposite direction, to the west, stars and everything else rising and setting past the horizon. As the Earth rotates around its poles, the sky appears to rotate about a pair of analogous *celestial poles* that lie directly above them. (Stand at the north pole, and the north celestial pole is above your head). Above the Earth's equator (the circle halfway between the poles) is the *celestial equator*. If you are in the northern hemisphere, you will see the north celestial pole above the northern horizon (by an angle equal to your latitude); if you are in the southern hemisphere, the south celestial pole will be above your southern horizon. Everywhere on Earth except at the poles the celestial equator runs across the sky from exact east to exact west, its maximum height above the horizon also depending on latitude.

As the Earth revolves around the Sun, the Sun appears to follow an imaginary path around the Earth (the *ecliptic)*, taking a year to make the circuit. The Earth orbits in the same direction in which it rotates on its axis, which makes the Sun appear to drift slowly eastward (a degree per day) relative to the background stars, counter to the direction of its daily motion across the blue sky. The Earth's rotation axis is not straight up and down, but tilted by 23.4 degrees from the orbital perpendicular. Because of the axial tilt, the ecliptic is inclined to the celestial equator by the same 23.4 degrees. As a result, the Sun moves back and forth, from 23.4 degrees north of the celestial equator to 23.4 degrees south of it, going from high in the sky at some times of the year to low at others. When the Sun is high, the Earth's surface absorbs more of its heat. When the Sun is low, the ground absorbs less, and it turns cold. The tilt of the Earth's axis is the sole cause of the seasons.

About March 20, the Sun crosses the celestial equator going north. The southerly crossing takes place about September 23. At these times, the Sun rises due east, sets due west, and is up for as long as it is down, these two points thus called the *equinoxes*. Passage of the Sun in March across the *vernal equinox* marks the beginning of northern hemisphere spring, while the *autumnal equinox* passage in September signals the start of autumn. (The points of maximum separation from celestial equator are the *solstices*; the northern, or summer, solstice is in Gemini, the southern, or winter, in Sagittarius, these parochially named by northerners.)

We find our way on Earth by means of a grid of latitude and longitude. Latitude measures position north and south of the equator, while longitude measures it east and west of a *prime meridian* that runs through Greenwich, England (because the English beat the French in a war). Astronomers use the same kind of grid in the sky, though with different terminology. In the sky, latitude becomes *declination* (that of the celestial equator 0 degrees, that of the celestial poles 90 degrees north and south). Longitude becomes *right ascension* (there is no "left"), which is always measured eastward from the vernal equinox. Declination can be measured with protractor-like devices (modern ones fitted with telescopes). Right ascension can be measured with a clock. The astronomer notes the time at which the vernal equinox crosses the *celestial meridian* (the sky's north-south line) and then notes the crossing time of the star. The right ascension (usually given in time units, where 1 hour equals 15 degrees) is the difference between the two.

In ancient Greek times, right ascensions and declinations of stars were measured with the naked eye. In the second century B.C., the great astronomer Hipparchus could determine positions to a fraction of a degree, and in the pre-telescopic sixteenth century, Tycho Brahe achieved a minute of arc (1/60 degree). At the start of the twentieth century, with telescopes, we could measure to 1/10 second of arc (a second 1/60 minute). Satellites have pushed the limit to under 1/1000 second, the angle made by the width of a penny in Los Angeles as seen from New York. And under special circumstances, relative angles can be measured to an accuracy 10 times greater.

There are more stars in the sky than anyone could or would want to count. Of these, astronomers have measured the positions of millions. We know where they are so well that we can watch them move relative to each other and can predict how the constellation patterns will be changed millions of years hence. More importantly, we can learn how the stars orbit within our Galaxy.

In contrast, the third dimension is difficult to explore. Astronomers have developed many ways of finding stellar distances, some born of sheer desperation. The fundamental method, on which most others ultimately depend, is *parallax*, the apparent displacement of an observed object caused by a change in the observer's position. You see in three dimensions because your two eyes look at a nearby object from two directions. Expand your head to the size of the Earth's orbit and you might see stars in 3-D. As the Earth orbits

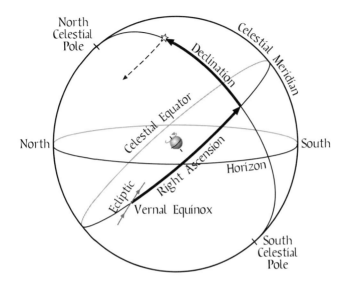

Imagine yourself standing on "top" of the Earth at the center of the picture, its seeming orientation dependent on the observer's latitude (here 50°N). The sky appears as a surrounding sphere. As the Earth rotates around its poles, the sky seems to rotate in the opposite direction around the celestial poles, and all the stars seem to move past the celestial meridian in daily paths parallel to the celestial equator, their rising and setting points depending on their declinations (angles from the equator). The Sun crosses the equator at the vernal equinox each year on March 20 or 21, the point providing the reference for right ascension, a coordinate analogous to terrestrial longitude, which with declination provides exact stellar locations.

and moves from one place to another, you achieve the same effect: the nearby stars appear to shift back and forth against the background of more-distant ones. The farther they are, the smaller the apparent shift.

But even over the Earth's orbit, the parallax shift is still unnoticeable to the naked eye, and its seeming absence was used in ancient times as evidence that the Earth did not move. When in the sixteenth century Nicolaus Copernicus firmly established that the Earth did so move, that it orbited the Sun, the absence of stellar parallax was taken to show that the stars are far away. The first parallax was detected telescopically only in 1838, when Friedrich Bessel, the renowned German astronomer and mathematician, watched the star 61 Cygni move over the year by 1/3 second of arc. Astronomers later formalized the word *parallax* to mean half the shift. The distance, expressed in *parsecs* (for *pa*rallax and *sec*ond), is just the inverse of the parallax, where the parsec equals 206,265 AU and is 3.26 times longer than the more familiar light year. Alpha Centauri has a parallax of 0.742 seconds of arc, rendering it 1.35 parsecs, or 4.4 light years, away.

Until 1998, decent parallax distances were limited to around 100 light years, far too short of what was needed to encompass the huge variety of stars that Nature offers. The Hipparcos satellite (after Hipparchus) revolutionized the game by measuring the parallaxes of hundreds of thousands of stars with at least fair accuracy out to distances of 1000 light years. These measurements provide an excellent basis for calibrating other methods. We are finally making good maps of our local surroundings. We know where the stars are, and from their locations where *we* are.

measuring the stars

Location does not by itself serve to tell the stars' real natures. We need physical measures, beginning with luminosity and temperature. Hipparchus divided the stars into six brightness categories called *magnitudes;* first magnitude contains the sky's brightest stars (21 of them, not counting the Sun), sixth magnitude the faintest he could see. To the eternal frustration of physicists (who would prefer some kind of standard physical unit), the system has survived the past 2000 years. In response to the logarithmic nature of human vision, magnitudes are logarithmic too: Hipparchus's first magnitude was roughly 100 times brighter than his sixth. Modern astronomers quantified the system so that magnitude 1.00 was exactly 100 times brighter than magnitude 6.00, thus sending some of the "first" magnitude stars (and planets) into lower numbers (including zero and even negative numbers). Sirius, the sky's brightest star, is of the minus first magnitude (−146); Venus at maximum is −4, the full Moon −12. Each magnitude division is the fifth root of 100—2.512—times brighter than the next fainter one. Telescopes allow the measure of progressively dimmer stars and galaxies to magnitude 30, a trillion times fainter than magnitude zero.

These magnitudes report only *apparent* brightness as seen with the human eye. A star (such as zeroth-magnitude Alpha Centauri) might be apparently bright because it is nearby; or it might be very luminous in total energy output, in watts, but apparently faint because it is far away. So (further frustrating physicists) astronomers adopted a system of *absolute* magnitudes (M), arbitrarily defined as what the apparent magnitude (m) *would* be were the star at a distance of 10 parsecs, or 32.6 light years. Once the distance of a star is measured by parallax, the calculation of absolute magnitude is easy. (For those who enjoy equations, $M = m + 5 - 5 \log D$, where D is the distance in parsecs. Given m and D, M follows quickly; moreover, if we can guess M and measure m, we have another way of finding distance.)

Absolute magnitudes reveal the immense differences among stars. At the standard distance of 32.6 light years, the Sun, at $M = +4.8$, would shrink to about the same brightness as the fainter stars of the Little Dipper, and could not be seen from town. On the other hand, Deneb, at the top of the Northern Cross, would glow 15 times brighter than Venus at her brightest. Absolute magnitudes of stars range from -10, a million times brighter than the Sun (three sets of five magnitudes, or a factor of $100 \times 100 \times 100 = 1,000,000$) to $+20$, a million times fainter. It would take a trillion of the faintest observed stars to equal the luminosity of the brightest.

Color complicates the issue. Stars take on noticeable, though subtle, tints. The degree of color is personal, different eyes seeing different shades: what one person sees as orange another sees as yellow. When passed through a prism, sunlight spreads into a lovely array—a spectrum—of colors, with red at one end, violet at the other. The solar spectrum shows that the yellow-white light of the Sun is really made of all the colors of the rainbow, the natural spectrum made when sunlight passes through, and is refracted by, falling raindrops. The amount of light radiated in each color by a star depends on temperature. The Sun, at 6000 degrees Kelvin, produces most of its light in the yellow-green, just where the human eye is most sensitive (Darwin at work). The combination of colors makes sunlight a bit yellowish to the eye. Cool stars near 3000 Kelvin or below, which appear reddish or orange, radiate most of their light in the low-energy red (and longer-wave infrared) part of the spectrum, whereas hot stars, say 20,000 to 50,000 Kelvin, radiate all across the colored spectrum and relatively much more in the high-energy blue and violet (not to mention the ultraviolet).

Cool reddish stars and hot bluish ones radiate best in the parts of the spectrum where the eye is least sensitive. A red-sensitive detector will quickly pick out red stars that the eye sees as faint. Imagine a star so cool that it radiates only in the infrared: whereas an infrared detector might see it as first magnitude, the eye would not see it at all. Hot blue stars suffer a similar problem. Consequently, different kinds of "eyes" see stars with different brightnesses, resulting in a proliferation of magnitude systems for different regions of the spectrum, from the ultraviolet into the far infrared. Magnitudes as seen through the human eye therefore become *visual magnitudes*, to distinguish them from the others. Because visual magnitudes describe only the "yellow-green" brightness, astronomers have to correct absolute visual magnitudes for unseen radiation to determine a star's luminosity output in watts, easily done once the temperature is known, a temperature that in turn can be found by comparing brightnesses made with detectors sensitive to different spectral colors. We are on our way to knowing the stars. To know them better, look at their social lives.

duplicity

When employed as an example of a typical star, our Sun has one great failing. It is single. Ignore the planets; the Sun has no stellar companion. Many stars do have mates, each in orbit about the other. Like people, if close enough, they can influence each other in profound ways, rendering the Sun as a typical stellar example somewhat misleading. We must therefore take the pairs and study them as units. Doing so also allows us to learn much more about stars in general, about what underlies all stellar differences. Stellar duplicity thus feeds back into our understanding of the single Sun.

To imagine the Sun a double—a *binary*—star, go no farther than our nearest neighbor, Alpha Centauri. A brief look through an amateur's telescope shows not one star, but two. The brighter, of apparent magnitude zero, is almost the same temperature as the Sun, but a bit larger and 60 percent more luminous; the fainter (first apparent magnitude) is a bit cooler than the Sun and shines with about half a solar luminosity. As the odometer turned to the year 2000, they were separated by 16 seconds of arc (0.0044 degree). Over the years they spiral about each other, drawing to a tenth that separation, then to half again as great, returning to the present angular distance after a slow dance of 80 years.

Bodies orbit in response to mutual gravity, each of a duo held in the grip of the other, each moving about a common point between them. The rules were laid out by Johannes Kepler and Isaac Newton over 300 years ago. (Kepler found the basic laws of planetary motion directly from Tycho's observations of the movement of Mars; Newton showed Kepler's laws to be a natural response to gravity and expanded upon them.) Closed orbits are elliptical (the circle a special case of the ellipse). The common point about which they move is proportionally closer to the more massive body. As the Moon orbits the Earth, the Earth orbits the Moon, but since the Moon is 80 times less massive than Earth, the common point is 80 times closer to (and actually inside) Earth. The massive Sun dominates the planetary system and hardly moves at all (about by its own radius in response mostly to Jupiter), so for all practical purposes we say the planets orbit it.

The farther apart the bodies, the weaker the mutual attractive force; the weaker the force, the slower the bodies move, resulting in a longer orbital period. The Earth, at 1 AU, takes a year to orbit the Sun, but Jupiter, at 5 AU, takes 12. The orbital period also depends on the sum of the masses of the two bodies: double the mass of the Sun and at their present distances, the Earth would speed around it in 8 1/2 months, Jupiter in 8 1/2 years. From the equations that govern orbits, the sum of masses of the Earth and the Sun can be derived from the orbital size and period [(period)^2 = (constant)×(size of orbit)^3/(sum of masses)]. Since the period and the orbital size of the Earth are known, and because the Sun is so much more massive than Earth, we can find one of the most fundamental of all astronomical numbers, the solar mass of 4×10^{33} grams, 333,000 times that of Earth.

Unlike the planets, the stars of Alpha Centauri have highly elliptical orbits about each other that take them from 24 AU apart (almost as far as Neptune is from the Sun) to 11 AU. The planets orbit our Sun, perpetually passing against the constellations of the zodiac, sometimes seen projected toward the Sun, sometimes away, in the opposite direction. If we could transport our Earth to orbit the brighter component of Alpha Centauri as it does the Sun, we would see the fainter component as a brilliant star that would take 80 years to go around our new Centaurian zodiac. When lined up with the brighter star, as it would be once a year, it would have little impact, but when seen at night it would on average light the sky as bright as thousands of full moons. Imagine the mythologies that this planet's societies would have developed before they learned what the system was all about.

Alpha Centauri is hardly unique. Some 20 percent of stars examined with a small telescope also appear double. A few are so widely separated you can see the individuals with the naked eye. Look at Mizar, the second star in from the end of the handle of the Big Dipper, and find dimmer Alcor 0.2 degree from it. At about the same distance from us and moving through space together, the two must take over a million years to orbit each other. In contrast, some pairs are so close that the individuals can barely be discerned. (As a result of twinkling, caused by refraction in the Earth's turbulent atmosphere, without very sophisticated equipment the practical limit for visible separation is about half a second of arc; the orbiting Hubble, whose vision is unaffected by the air, can do ten times better). If close pairs are of different temperatures, their proximity enhances their contrasting colors and the doubles become startlingly beautiful. Nineteenth century astronomers tried to outdo each other in description (azure, gray, dusty blue, and so on).

Close but still visually separable pairs have orbital periods within human lifetimes; some are even as short as a few years. Accurate orbits have been measured for hundreds of binaries. The orbital equations give the sum of the masses of a double's two stars. Careful observation also locates the common center, revealing the ratio of masses; the sum and the ratio together yield the individual masses. We can thus weigh stars through their duplicity. Directly observed masses range from about a tenth that of the Sun to more than 20 times solar.

Masses can also be correlated against luminosities. The true brightnesses of stars like the Sun, those that are running on hydrogen fusion in their cores, increase very quickly with mass. Double the mass, and the star becomes 10 times brighter. The more mass a star has, the greater the internal compression and temperature, and the greater the luminosity should be, theory nicely matching observation. From this relation, the most luminous stars must be over 100 times heavier than the Sun. Mass is a star's single most important parameter, governing not only its luminosity, but its entire career!

If the binary components are close enough, they merge into one telescopic image. But the individuals still reveal themselves. Spread a star's colored spectrum with enough clarity, and find it broken by myriad narrow gaps where light appears to be cut out. Each of these gaps is produced in the star by a specific kind of atom that absorbs the outflowing light, as identified from laboratory studies. The gaps—called *absorption lines*—are used to analyze the star's chemical makeup, but they can also be used to detect close doubles.

Move toward a wave, and its frequency (the rate at which the wave peaks hit you) seems higher, the wavelength shorter; move along with a wave and see the reverse. This *Doppler effect* works with any relative motion along the line of sight. If a car comes at you, the sound waves it generates seem higher, and vice versa. If a star moves toward you, its light waves are shifted to higher frequencies and shorter wavelengths, and vice versa. The size of the effect depends on the speeds of motion relative to the speed of the wave, so the Doppler effect is obvious in slower sound and subtle in much faster light [(relative wavelength shift) = (velocity of body)/(velocity of wave)]. But tiny shifts in the measured wavelengths of the absorption lines give the effect away. From the degree of shift we can find the speed of a star along the line of sight. When stars orbit each other, they are continuously changing their line-of-sight speeds relative to Earth. Therefore, the absorption lines periodically shift back and forth, allowing us to detect and analyze pairs that we cannot actually separate at the telescope. Similar measures, though with different rules, allow the study of the expanding Universe.

If the two stars of one of these very close pairs are about the same brightness, we see two sets of absorptions shifting back and forth in wavelength as the stars circle each other and change their line-of-sight speeds. The determined orbital speeds allow us to find the orbital sizes and again masses, even though the stars are not individually visible. Now we can study stars that are so close together that they orbit not in years, but in days, some even in hours, stars much closer together than Mercury is to the Sun, others so neighborly that they are nearly in contact and distort each other through mutual

wavelength (Å)

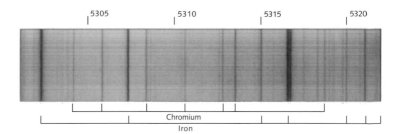

Stretch sunlight—or any starlight—into its colored
spectrum, violet at one end, red at the other, and
see it crossed by thousands of dark gaps, each
produced by a specific atom or ion. This very small
"yellow" section of the solar spectrum reveals
hosts of gaps—*absorption lines*—caused by iron,
chromium, and some rarer elements. The numbers
across the top are the wavelengths of the light in
Ångstroms, a unit equal to 1 hundred-millionth of
a centimeter. (You could not sense the color difference from one end of this spectrum to the other.)
Almost nothing can be learned of a star without
analysis of its spectrum. (Mt. Wilson and Las
Campanas Observatories, Carnegie Institution of
Washington, courtesy of E. C. Olson.)

tides. (The gravitational force of one body on another is stronger on the near side and weaker on the far side, resulting in that other body's being stretched. The Moon stretches the solid Earth on the line that connects the two and, since water flows easily, particularly stretches the oceans. Tides raised by the Earth in the solid body of the Moon have caused the Moon to face the Earth perpetually.) At the most extreme, one star can even destroy the other.

But now, that many-headed Hydra of astronomy—the third dimension—rises to impede our journey toward stellar truth. If the orbit is set perpendicular to the line of sight, there will be no changing line-of-sight velocities and no Doppler shifts. So what we actually get from these measures are lower limits to the velocities, orbital sizes, and therefore masses. But as Hercules slew Hydra, so orientation slays uncertainty. If the plane of the orbit happens to be close to being *in* the line of sight, the circling stars will block—eclipse—each other's light, and the pair, seen as a single star, will dim. The most famous example is Algol—the "Demon Star"—in Perseus, which dims by over a magnitude (second to third) every 2.87 days when a small bright star passes behind a much larger one. The eclipses are so regular you could keep time by them for the next millennium.

Hundreds of such stars are known and studied. From the way the light drops in eclipse, we can find the actual orbital tilts, and if Doppler shifts are present, find the true masses. The durations of the eclipses also give the sizes of the stars. The year-long eclipse of a small bright star in the VV Cephei system tells of a star the size of the orbit of Saturn.

Double stars are hardly the story's end. The Alpha Centauri system is actually *triple*. The bright pair is orbited at a distance of roughly 15,000 AU by much fainter Proxima Centauri, which is currently the closest actual star. It is so faint that seen from a hypothetical Earth orbiting Alpha Centauri's bright component, Proxima would be only fifth magnitude, not that much brighter than Uranus is in our home sky. Mizar is better yet. Orbited by dimmer Alcor, Mizar is itself double, its two components 14 seconds of arc apart. And spectra show that each one of the Mizar pairs is yet again double, making Mizar a double-double star. With Alcor, the system is quintuple. The classic double-double is Epsilon Lyrae, the pairs visible with excellent eyes; each clearly splits again visually through the telescope. Castor, in Gemini, is a close double-double orbited by a distant double, a sextuple star! Obviously nature likes to double up. Our Sun, in contrast, rules the Solar System in lonely and somewhat unusual splendor.

congregations

In northern autumn the lovely compact Pleiades, the Seven Sisters, surmount the horizon (most eyes see six stars, some eight). In mythology they are the daughters of Atlas and also nymphs of Diana pursued by Orion. They are the epitome of the *open clusters* that dot the Milky Way. Hundreds are known and thousands inhabit the Galaxy. Unlike structured double-doubles, their stars exist in congregation, all bound together by gravity. Many hundreds lie within a volume typically 10 or so light years across, and all members orbit a common center of mass. Within the open clusters we find the same kinds of doubles as among non-cluster stars.

Several open clusters are near enough to be visible to the naked eye; the Pleiades is only 425 light years away. The closest obvious cluster is the Hyades. At a distance of 150 light years, it appears quite large and makes up the entire head of Taurus the Bull; the bright star Aldebaran, not a member, is positioned in front of it. At 260 light years (but without the bright stars that make the Pleiades so promi-

The "Jewel Box," which abuts the Southern
Cross, is a wonderful example of an open cluster,
a loose collection of stars ten or so light years
across that will eventually break up. Such clusters
abound in the Milky Way, showing that stars are
commonly formed in groups. Perhaps the Sun
once had such a family. (© Anglo-Australian
Observatory, Photograph by David Malin.)

nent) lies lacy Coma Berenices, which forms a constellation below the handle of the Big Dipper. Deeper examination reveals the Beehive in Cancer, the Jewel Box next to the Southern Cross, M7 in Scorpius ("M" for Charles Messier, who in the late 1700s catalogued bright clusters and other celestial objects), and numerous others. Hardly recognizable is the Ursa Major cluster, which consists of the middle five stars of the Big Dipper (averaging 80 light years away) and many others. There are so many open clusters that we can reasonably assume that stars are born within them.

Though seemingly large and invulnerable, on a Galactic time scale most open clusters do not last very long, only a billion or two years. Gravitational interactions among the members of an open cluster can speed up a star so much that it is hurled free, the cluster thus reduced. Tides raised by the Galaxy stretch the cluster and have the same effect of ejecting the outer, more loosely bound, stars. Only those clusters found in the fringes of the Galaxy, where gravitational pulls are weak, survive to an old age of several billion years. It is not inconceivable that the Sun was born in a cluster, its siblings long since lost. The existence of so many clusters in spite of their falling apart is evidence that clusters and their stars are continuously being born to replace those that die.

As beautiful as they are, open clusters, which occupy the disk of our Galaxy, pale beside the much rarer *globular clusters* that lie outside the disk in a grand surrounding halo. Open clusters are lovely; globular clusters are both lovely and spectacular. Three are visible to the naked eye, M13 in the constellation Hercules in the northern hemisphere, and two (47 Tucanae and Omega Centauri) in the southern. A small telescope reveals dozens. In the most glorious of them, over a million stars are packed into a volume no more than 100 or so light years across, most tightly compacted to the center.

Contrasting mightily with open clusters, globular
clusters—like Messier 80 in Scorpius—are made
of vastly more stars, up to millions, and are so
tightly compacted that they can survive almost
indefinitely. Vastly older than open clusters,
globulars go back to the formation of our Galaxy
itself, before it developed the disk we now see.
(Space Telescope Science Institute, Hubble
Heritage Team; AURA/StScI/NASA.)

Even the smallest globular cluster has thousands of stars. Their failing is their rarity. Only about 150 are known in the Galaxy, but given their extraordinary natures, who cares?

Globular clusters have so many stars and are so tightly compacted that most, if not all, survive the onslaught of the gravitational forces that cause stellar escape. If you lived at the center of one, you would see thousands of first magnitude stars in your sky. They are all very old, and date back to the time of the beginning of the Galaxy itself. Indeed, they must contain profound clues to the Galaxy's origins.

home

Single stars, doubles, quadruples, and clusters of both kinds all assemble into the next stage in the hierarchy, the *galaxy*. We cannot see our own Galaxy very easily because we are inside it. Our best initial view comes from galaxies that might be like our own. Once we know what to look for we can refine the view of ourselves and then find still other galaxies that match even better. A century of research has yielded a picture of a complex system.

The deceptively simple textbook view of our Galaxy is that of a thin disk some 80,000 light years across surrounded by an enormous spherical halo. The disk contains 98 percent of the Galaxy's 200 billion stars, the vast bulk of interstellar matter, and all the open clusters. Though the halo is sparsely populated with stars, it contains the collection of massive globular clusters. The stellar populations of both the halo and disk increase toward the center, where they merge to produce a vast bulge in the disk. We live in the disk 25,000 light years, about two-thirds the radius, from the center. We see the

Though galaxies come in many forms, the grand-
est is certainly the *spiral*, in which a flat disk of
billions of stars breaks into graceful spiral arms
that wind away from a dense nucleus. If this one
(NGC 4603, depicted by Hubble) were our own,
our Sun would be just inside one of the outer
arms. From our vantage point we would see this
disk as a "milky way" around our heads. Open
clusters occupy the densely-packed disk, while
globular clusters, as well as individual stars,
occupy a sparse surrounding halo. (Jeffrey
Newman, University of California at Berkeley;
AURA/StScI/NASA.)

combined light of the disk at night around us as the thick band of the Milky Way, which varies considerably in brightness around its circle because of our off-center position.

The stars within our Galaxy must be in orbit around its center; were they not, the whole mass would fall into a dense lump under their combined gravitational contraction, and stars could not exist. The orbital period of a star depends on the amount of mass in the Galaxy interior to its orbit, the star and inner Galaxy acting as "two bodies." Since stars are at different distances from the center, all orbits are different. (And even if the distances were alike, the orbital shapes would not be quite the same.) The orbiting stars therefore shift position relative to each other, dooming the familiar constellations. By the time the Sun is on the other side of the Galaxy, hardly any of the stars you see tonight will be visible, and a new set will define new constellations.

Over only a few years these motions are detectable by relative changes in stellar coordinates. Combined with distances, they let us obtain speeds across the line of sight. The Doppler effect yields radial velocities *along* the line of sight. The result is full knowledge of a star's speed and direction relative to the Sun, the disk's stars typically drifting by us at 20 or so kilometers per second. Doppler measurements made of globular clusters and other nearby galaxies outside the disk allow the determination of the solar orbit. We speed around the Galaxy's center on a nearly circular orbit at 220 kilometers per second, taking 230 million years to make the circuit. Since the Sun and Earth are 4.5 billion years old, they have gone around 20 times! (Our neighbors are temporary indeed!) From the movements of the other stars of the disk relative to the Sun, we find them to trace more or less circular Galactic paths as well, with periods increasing outward from the center. The stars of the halo are distinctly different, as their orbits are highly elliptical. As halo stars crash past us through the disk they speed by at up to hundreds of kilometers per second.

Pictures of other disk galaxies show graceful spiral arms pinwheeling out from the center, making the galaxies look as if they are rotating. When we examine our own Galaxy, the distribution of stars, clusters, and interstellar gas reveals that ours, too, has spiral arms. The Sun is on the inner edge of one. Their effect can even be seen in variations in the brightness of the Milky Way. The spiral arms are actually gravitational disturbances propagating through the Galaxy, stars and interstellar matter piling up in them. The arms are (over millions of years) temporary. New ones replace those that dissolve, and disk stars like the Sun move in, then out, of them. As repositories of dusty interstellar gas, the spiral arms are the factories that create stars, which in turn outline the arms.

But as motion and gravity dissolve spiral arms, reality dissolves simplicity. There is no Galactic "edge." The sharp-lined textbook view shows only the boundary that contains 90 percent of the stars. The population of the disk and halo continue, though with greatly decreasing density, well beyond this arbitrary limit. The disk has been detected to over 80,000 light years from the center, twice the distance of the purported edge, and it certainly extends farther. Globular clusters are seen to over 100,000 light years away from the center. The disk is also highly structured; interstellar dust and gas run down the middle in a thin disk only a few hundred light years thick. Different kinds of stars constitute thicker disks, much like a layered bun around a hamburger. The halo is structured as well, one set of globulars making a very thick disk embedded in the halo. The evidence is consistent with our Galaxy having formed through gravitational collapse of a pre-galactic blob of matter, punctuated by a series of collisions with other galaxies that resulted in both mergers and tidal disruptions. Our Galaxy, it seems, is a mess.

And so is our conception of it. Much of its structure eludes us. It is so dusty that we cannot see very far within the disk, though radio telescopes, which detect interstellar gas, can readily puncture the haze. Like many other spiral disk galaxies, ours seems to have a bar running through the center from which the arms emerge, but we are not at all sure of the bar's structure. More dismaying, since most of the stellar mass of the Galaxy is concentrated toward its center, we would expect stellar orbital speeds to fall off toward the outside, much as Jupiter orbits the Sun more slowly than Earth. They do not! Continuing high orbital speeds show that larger orbits encircle progressively larger amounts of gravitating matter. But we cannot see it: dark matter revealed! Unfortunately, nobody knows what the dark matter might be. But we do know from gravitational analyses that it constitutes as much as 90 percent of the Galaxy's mass and that it forms a "dark matter halo" ten times the size of the canonical "edge." The illuminated Galaxy is but a core set within a massive system of mysterious matter. Motions within other galaxies show that ours is hardly unique. Dark matter dominates the Universe.

everybody else

Galaxies flock everywhere around us, galaxies of myriad forms and sizes, no two quite alike, some wildly different from ours. The nearest one comparable to ours can be seen with the naked eye as a fuzzy patch in Andromeda: the Andromeda Galaxy, M31. Though 2 million light years away, it shines at fourth magnitude and is the most distant

thing the human eye can see without a telescope. Even more magnif-
icent than ours, with about twice the number of stars, its spiral arms
and dominating central bulge are obvious in a system pitched to the
line of sight by about 70 degrees. Not far away, in Triangulum, is
M33—a smaller, though no less lovely, spiral seen more face-on.

These two galaxies along with ours dominate the Local Group,
our very own cluster of about three dozen galaxies. All are bound
under the force of gravity and all orbit a common center of mass, the
system stable against the expansion of the Universe. The three serve
as centerpieces for yet smaller systems. To our Galaxy belong the two
Magellanic Clouds. Appearing as fuzzy patches of light, and visible
only below about 20 degrees north latitude, their name honors the
world explorer. About 150,000 light years away and about 1 percent
our mass, the Large Magellanic Cloud seems to try desperately hard
to become a spiral but cannot quite make it. Star formation, however,
is rampant, far greater than in the neighboring unstructured Small
Magellanic Cloud. M31 likewise dominates some small elliptical
galaxies (including prominent M32), systems that have ellipsoidal
shapes and no spiral arms or much in the way of star formation or
interstellar matter. At a lower level are small scraps and shards of
galaxies, some associated with the "big guys," others not, some pulled
by tides from the others, all hunting down mates with which to merge.

Our small Local Group is near the edge of the massive Virgo
Cluster, which contains more than 2000 galaxies centered 50 million
light years away. The system is dominated by a massive elliptical
galaxy (M87) over 10 times the mass of our own. Now we begin to
see the Universe's expansion, as our Local Group and the Virgo cluster are flowing away from each other. Still, they are close enough for
gravity to have a significant influence, as the observed flow rate is not
quite as fast as would be expected from the overall expansion rate. In

Galaxies love company, and tend to clump into
groups and clusters, clearly a clue to how they
were born. This distant cluster contains thousands.
(Alan Dressler, Carnegie Institution; and NASA.)

astronomical parlance, we are "falling" toward the Virgo Cluster—but shall never arrive.

Beyond Virgo are thousands, millions of other clusters, big ones, small ones, single galaxies spattered between them, all adding up to a trillion or more observable systems. They are arranged not just in clusters but in bigger organizations of superclusters, and collected into enormous walls, the whole having something of a sponge-like structure. And here we return to where we began, these assemblies having come from small perturbations in the Big Bang.

All is now in place; let us return to the stars.

the family

The members of a human family are related, but that does not mean your Uncle Ed looks or behaves anything like your mother. So it is with stars. All stars are related through a variety of physical processes, yet all are different, some amazingly so, rather like Mom being a ballerina and Ed a sumo wrestler (as unlikely as that may be). Even as identical twins have subtle differences, no two similar-appearing stars are really quite alike either.

As we would come to know any extended human family, to understand the family of stars we must first examine its members and see what subgroups there may be. To understand any of these, whether individual stars or their immediate families, look to the star we see best, the one we see close up, the one we observe in exquisite detail, the Sun.

our very own star

The Sun is too bright to look at without proper protection. Do not try to view directly what is described here. Professional equipment and proper training are essential. But how unfortunate it is that we can't just look at the sun, because there is so much to see. The solar surface—the part of the Sun where the solar gases suddenly become opaque —seethes with motion and energy. A gaseous sphere over 100 Earths across, containing a mass 330,000 times our Earth's, the Sun's 6000 degree Kelvin surface pours 4×10^{26} watts of radiant energy into space, only half a billionth of it caught by our planet.

Filter the brilliant sunlight and zoom in on the solar surface. Instead of the smoothness it presents when seen through a misty haze, it breaks into a million bright grains separated by dark lanes, each bright patch the size of a smaller European country. Even as we watch, the grains dissolve, new ones taking their places within five minutes or so, the Sun a boiling cauldron. Most energy in the Universe is moved by radiation, but if conditions are right, mass motions, the gross movement of matter, can also do the job. The Sun's temperature increases inward, from 6000 degrees Kelvin at the surface to 15 million at the center. The rate at which temperature changes with distance renders the gases in the outer third of the Sun unstable, making them subject to *convection*, in which hot gases rise and cool ones fall. The grains are the tops of giant convection cells, in which columns of hot gas bubble up from below and release their energy as radiation when they hit the top. The cooled gas then darkens and falls back, eventually picking up another load of heat energy from deep in the Sun.

Superimposed on the bright grains are larger dark regions, called *sunspots*. Some are just barely visible, while others may be much larger than Earth, so big at maximum that (with appropriate filtration) they can be seen with the naked eye. A few spots may be isolated, but most are found in pairs and massive complex groups. The spots look dark only in contrast to the rest of the solar surface, as their 4500 degree Kelvin gases radiate much less light than do the grains. Like the grains, spots come and go, but over longer time scales, some lasting a day, others a month or more. Each day, patiently count the spots. Over the years the number visible on the solar surface changes dramatically, growing slowly to a maximum, falling. The cycle repeats every 11 or so years, each cycle different from the one before.

To see what is going on, turn again to the spectrum, to the absorption lines, the gaps in the spectrum caused by individual atoms and ions (atoms with electrons missing from them) whose positions were earlier used to detect Doppler shifts. Each atom or ion has its own unique set. The Sun's absorption lines—tens of thousands of them—are primarily used to find the chemical composition of the solar *atmosphere*, the razor-thin, partially transparent layer from which sunlight escapes into space. The gaps are not all the same: some are powerful and broad, taking huge chunks from the rainbow of color; others are weak and barely detectable. If we compare the strength of an absorption made by iron with the strength of one made by hydrogen, and apply ample atomic theory, we can find the iron-to-hydrogen ratio, the number of atoms of iron relative to the number of atoms of hydrogen.

Full analysis shows the outer parts of the Sun (and because of the convection, the whole outer third at least) to be 92 percent hydrogen and (from other data) 8 percent helium. All the other kinds of atoms make up a mere 0.15 percent. These proportions are amazingly different from those of Earth, which is mostly iron, nickel, silicon, and oxygen, with hydrogen and helium relatively quite rare. However, if we set aside the Sun's two light elements (hydrogen and helium),

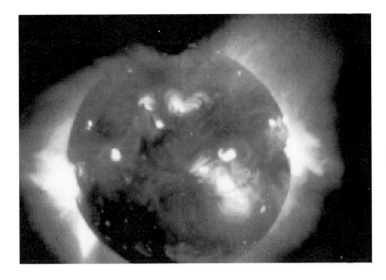

The Sun, while overall stable, is remarkably
active. Its rotation and outer convection
generate vast and powerful looping magnetic
fields that enter and exit the Sun at sunspots.
Magnetic energy heats the gas in the loops to
such high temperatures that it emits X-rays
(as seen here), giving the Sun a very different
appearance. The heated gas is seen by eye
during a solar eclipse as a surrounding white
corona. (Yohkoh Observatory, Soft X-ray
Telescope [SXT] prepared by the Lockheed
Palo Alto Research Laboratory, the National
Astronomical Observatory of Japan, and the
University of Tokyo with the support of
NASA and ISAS.)

the relative ratios of the remaining solar elements are strikingly similar to the ratios found in the Earth's rocky crust, demonstrating again that Earth and Sun are part of the same family and were born together from the same parent cloud of interstellar matter.

When we look at the absorptions in the spectrum of a sunspot, they are split, the effect (observed in the physics lab) immediately telling us that a magnetic field is acting on the atoms. The fields are immensely powerful, thousands of times that of Earth's. Moreover, the magnetic fields of each of a pair of spots are opposites (one "plus," the other "minus"). The most reasonable explanation is that concentrated "ropes" of magnetism extend out from the Sun in great loops; the dark spots are formed where the loops enter and exit the solar atmosphere. The magnetic field in a loop is so strong that it stops the upward convection, locally chilling the solar surface and forming the spots. The loops are unstable, and when they die away, the spots disappear too.

The spots move across the Sun in a stately fashion, showing the Sun to be rotating at its equator with a period of 25 days, but closer to 30 toward the poles. Unlike the solid-body Earth, the Sun rotates "differentially." Its gases move faster at the equator, but more slowly toward the poles, so they slide, or shear, past one another. The interior gases of the Sun are not just hot but ionized, the electrons stripped from the atoms. The movement of the ionized gases by convection and rotation combine to produce electric currents that in turn create global magnetism, giving the Sun a magnetic field much like that of Earth, with a north and south magnetic pole. The differential rotation also concentrates the field into the magnetic ropes, which when raised up and out by convection make the spots. As the sunspot cycle—really a magnetic cycle—continues, the fields become ever more complex, until after 11 or so years the whole thing breaks down and starts over.

The great solar luminosity and magnetism combine in ways still not well understood to drive matter from the solar surface, creating a "solar wind" that blows at a rate of about a tenth of a trillionth of a solar mass per year. Gas both streaming from the Sun and confined by magnetic loops is heated to a temperature of 2 million degrees Kelvin by the continued release of magnetic energy to form the vast but faint solar *corona* that is visible only during a solar eclipse or with specialized equipment. At first glance about double the solar radius in size, the corona really has no clearly defined limit as it extends outward toward the planets.

The mass-loss rate in the solar wind is too small to have much of an effect on the Sun, but it has a major impact on the Solar System. The solar wind blows back the ionized gases of comets to form their tails, and slams into the Earth to make the atmosphere faintly glow. Collapse of major magnetic field loops can send huge bubbles of coronal gases into the solar wind. Hitting the Earth, they cause the northern and southern lights, damage the electronics in orbiting satellites, and even cause major power outages. The Sun not only warms, but also reaches out to touch us! The power of the solar magnetic cycle was vividly demonstrated by a period in the late 1600s when it disappeared, and the Earth was plunged into the Little Ice Age, which drove out communities in Greenland and put northern Europe and North America into a deep freeze. Periods of warming and cooling tend to follow solar activity levels. We are still not sure why.

Here is our benchmark from which we can explore the stars. Do other stars exhibit the same phenomena? Given the solar luminosity and the known rate of hydrogen fusion, the Sun has enough hydrogen left in its core to last for another five or so billion years before it starts to die. What about other stars, in what states might they be? Can we see birthing ones and dying ones? How do binary companions affect the picture? And do other stars have planets from which someone might examine their own benchmark? Let's look.

by their letters shall ye know them

Like clearing your desk to begin a major project, the first step in stellar study is to organize. Three obvious stellar qualities stand out: luminosity, temperature, and mass. If we can find how they relate to one another, perhaps we can learn how stars live and die. Visual luminosity, or absolute visual magnitude, is calculated from apparent visual brightness and distance. Mass is determined by analysis of double stars, temperature through spectra. These three quantities provide the initial keys to modern understanding.

Spread starlight into a spectrum, and you see the usual absorption lines produced by atoms and ions within the star's thin atmospheric layers. Exploration of stellar spectra goes back to the middle of the nineteenth century. The early spectral pioneers were thoroughly confused, as some stars had spectra that were vastly different from others and from that of the Sun. Some stellar spectra are dominated by hydrogen, while others show strong effects of helium, ionized metals, neutral metals, and even molecules. The first step toward understanding the spectra was to place the stars into "bins," rather like categorizing trees as elms, maples, and sycamores.

Various taxonomic schemes culminated in one whose development began at Harvard around 1890. Very flexible, and all-pervasive in stellar astronomy, the classic system used seven simple main categories, plus a few side branches and extensions (and of course the odd bins for those stars that do not fit with the others). The system began its life by alphabetically ordering stars according the strengths of their hydrogen absorption lines. The astronomers who created it (William Pickering, the head of Harvard College Observatory, and his famed assistants, Annie Cannon, Williamina Fleming, and Antonia Maury) then dropped some letters as not needed and rearranged others to produce greater continuity for absorptions other than hydrogen. The result was the classic *spectral sequence*, OBAFGKM, to which L and T have recently been added at the end.

Nothing characterizes a star quite so much as its spectral class; it is the first thing about a star that any astronomer wants to know.

The Harvard astronomers quickly saw that the spectral sequence related strongly to stellar color. (Visual colors are mostly subtle; "color" for an astronomer is more related to where in the spectrum the energy is dominantly pouring out rather than to what is seen with the eye.) Color is also related to temperature. The higher the surface temperature, the more radiation is produced toward the more energetic violet end of the spectrum. The Sun, near 6000 degrees Kelvin, glows to the eye with a soft yellow light. But at 3000 degrees Kelvin, the stellar gases predominantly produce only lower-energy red, and infrared light. At 50,000 degrees Kelvin, on the other hand, so much high-energy radiation is produced that the star takes on a slightly bluish cast.

Now we can begin to describe stellar variety. Evidence for hydrogen is present in the vast majority of stars. But in the white class A stars (like Vega and Sirius), it is supreme. Spectral analysis shows them to have temperatures that fall between 9500 and 7000 degrees Kelvin. Below these are the yellow-white F stars such as Procyon in Canis Minor; with weaker hydrogen but stronger ionized metals, their temperatures range between 6000 and 7000 degrees Kelvin. Next in line are yellowish G stars, like the Sun. Their spectra have lots of metals, and temperatures fall in the 4500 to 6000 degree Kelvin range. Class K (Arcturus, Aldebaran), yellow-orange with neutral (un-ionized) metals, lies between 3500 and 4500 degrees Kelvin; and last in the classic sequence is reddish M (Betelgeuse, Antares), descending to 2000 degrees Kelvin. M stars are so cool that molecules, combinations of atoms so fragile that they are broken apart at high temperatures by atomic collisions, start to take over, and at the low limit dominate. The spectra of cooler M stars are filled with dense dark bands made by titanium oxide.

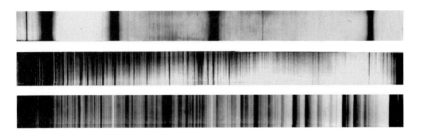

Even though most stars are made of the same
stuff, they possess a remarkable variety of
spectra. At the top is the blue spectrum of an
"A star," one with powerful absorptions of
hydrogen. In the middle is one of a "G star"
like the Sun, displaying weaker hydrogen and
a plethora of metal absorptions. The spectrum
at bottom has bands of titanium oxide characteristic of an M star. The amazing
differences are all caused by temperature,
from top to bottom 9000, 6000, and 3000
degrees Kelvin, which almost alone controls
the formation of molecules, the balance of
ionization, and the efficiency of absorption.
(University of Michigan Observatory.)

The new classes L and T extend downward to even lower temperatures. The L stars, only recently found with infrared detectors, drop to 1500 degrees Kelvin, a temperature so low that oxides exchange for hydrides (metals attached to hydrogen atoms), many of the metals also freezing onto solid molecular grains. At absolute bottom are the stars of tentative class T, near (even below) 1000 degrees Kelvin, which display absorptions of methane, reminding us of the spectra of the giant planets. At the other extreme, class A blends to blue-white hot B, hydrogen weakening, helium making its appearance in temperatures between 9000 and 25,000 degrees Kelvin. At the top, B extends to grand bluish class O with its ionized helium absorptions and temperatures to nearly 60,000 degrees Kelvin. Mighty Orion is filled with O and B stars, including Rigel, the belt stars, and many others.

However, a few broad groups for nearly all the stars proved to be insufficient. To allow for better discrimination, the system was decimalized. As temperature declines, class A is broken down from A0 through A9, which smoothly continues to F0 through F9, and so on. An A9 star has much more in common with an F0 star than it does an A0 star. The Sun's class is refined to G2, as is the brighter of the stars in the Alpha Centauri double. Sirius is A1, Vega A0, Procyon F5, Betelgeuse and Antares M1.

It originally looked as if all these stars were made of different chemicals. Only when physicists and astronomers understood how electrons bound to atoms and molecules could make the absorptions did they recognize that the differences among the classes concerned not chemical composition but ionization and temperature-dependent efficiencies of absorption. (Neutral hydrogen, for example, ionizes at hot temperatures and therefore weakens in classes B and O. It takes even more energy to ionize helium, so its spectrum appears only among the O stars.) In turn astronomers could then use the spectrum to derive compositions. Among the greatest twentieth century astronomical discoveries was that the stars of the classic sequence are all made of the same stuff. Like the Sun, they are about 90 percent hydrogen, 10 percent helium, and 0.15 percent everything else. Topping the "everything else" group is oxygen, followed by carbon, neon, and nitrogen, the rest way down the line. For every billion hydrogen atoms there are only 30,000 iron atoms, 40 of zinc, and 0.01 of gold. How very odd is the Earth, then, made almost entirely of these heavy materials.

giants and dwarfs

When scientists have two parameters to describe any physical system, they have an overwhelming urge to see if they correlate: does one increase as the other decreases, or vice versa, or is there no "signal" there at all? All measures have errors associated with them. Some of the greatest science has come from cleverly extracting correlations that are hidden by the "noise" of the errors; some of the worst by extracting correlations that were never there.

Nearly 100 years ago, two astronomers, Ejnar Hertzsprung in Denmark and Henry Norris Russell in the United States, independently looked at the growing body of stellar data and, like good scientists, correlated two sets of them, the graph of the two ultimately becoming known as the Hertzsprung-Russell, or HR, diagram. It plots visual luminosity upward (absolute visual magnitude perversely decreasing) and spectral class to the right, starting with O through M. (Since spectral class reflects temperature, the graph shows temperature perversely decreasing outward, annoying physicists yet more.)

For most stars, as the temperature increases (to the left) along the sequence from M to O, so does the luminosity. The correlation is quite understandable, since a hot body produces considerably more radiation than a cool one. However, the diagram contains a stunning surprise. There is a second and reverse correlation. Starting roughly at the position of the Sun on the diagram, there is a sprinkling of stars that, rather than becoming dimmer toward lower temperatures and class M, get *brighter*. Some stars with temperatures half that of the Sun are many times more luminous. What is going on?

A star's true luminosity—its power output—most certainly depends on its temperature. Double it and the energy radiated by each square meter of surface goes up by a factor of 16. That is the key; it is the radiance *per unit area* that depends on temperature. But nothing has been said about the *total* radiating area. For a star to be luminous though cool, it must have enormous surface area. Most stars, unless distorted by tides or severe rotation (and both happen), are spheres whose surface areas depend on the square of their radii (R). The luminance per unit area depends on temperature (T) raised to the fourth power. The total luminosity L is therefore a constant times T^4R^2. It takes a very large R to offset the power to which T is raised. The bright cool stars are not just large, they are huge.

Russell and Hertzsprung distinguished the two fundamentally different kinds of stars by overstating the case, logically calling the rarer big ones *giants*. But they called the more populous and smaller stars *dwarfs*, implying that they are somehow smaller than "normal." These dwarfs, however, are in fact the typical sorts of stars that populate the Galaxy. The Sun, as normal as can be and over 100 Earths across, is a dwarf! But only by comparison with the bigger giants. Even with the earliest data, giants were seen to be 50 times the solar size, half an astronomical unit (AU) across. If placed at the Sun, such a star would extend over halfway to the orbit of the planet Mercury.

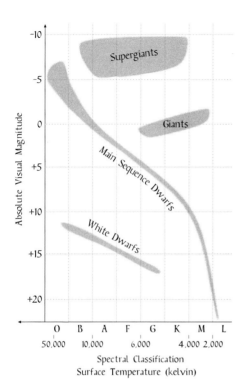

Stars are best studied by arranging them in a graph of luminosity plotted against temperature, classically done by using absolute visual magnitude (brightness increasing upward) and spectral class (temperature increasing leftward). The stars at the top are a million times the luminosity of the Sun, those at the bottom a millionth. Stars separate into ordinary hydrogen-fusing main-sequence "dwarfs" like the Sun, huge giants, yet-larger supergiants, and tiny white dwarfs, each kind in a different stage of life and death.

Some of the coolest giants are five or more times larger; they would swallow the Earth and extend to the orbit of Mars.

But wait: Hertzsprung had actually discovered stars that were more luminous than the giants. Many are just as cool and must therefore dwarf the giants! What do you call a star when a superlative term has already been used? Make way for the *supergiants*. Supergiants start where the giants leave off. The largest known (far brighter than giants, while also cool), are close to 20 AU across, 2000 times the size of the Sun, approximating the size of the orbit of Saturn. As the Sun could contain a million Earths, such a star could gorge itself on nearly 10 billion Suns.

When we take into account that stars have a variety of temperatures, the terminology grows quite literally more colorful. Giants and supergiants extend across the classic classes (ultra-cool L and T are confined to dwarfs). Thus astronomers speak of yellow and red giants, blue supergiants, red dwarfs, and so on, the color a quick reference to temperature. Since the HR diagram is numerically dominated by the common dwarfs, which diagonally slice the diagram from hot blue class O through cool red M (and even to deep red L and T), dwarfs have become more logically known as *main sequence stars* (though the dwarf appellation is still commonly used, the terms synonymous). Anthropomorphism remains undefeated, however. Not only are giants and supergiants still called such, but there is a smattering of stars that fall between the main sequence and the giants. For a given temperature, lesser in size than giants but bigger than dwarfs are the *subgiants* (better at least than "superdwarfs").

The modern HR diagram comes in many forms. In its classic inception it plotted—and still plots—absolute visual magnitude against spectral class. Spectral classes, however, can be laborious to obtain. Star color, which relates strongly to class, is easier to find (by measuring magnitudes with detectors sensitive to different parts of

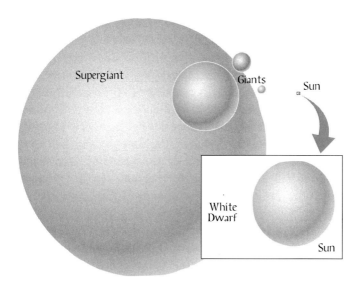

The range of stellar sizes is astonishing, the
biggest supergiants a hundred thousand times
the diameter of a white dwarf.

the spectrum), so it is also common to see absolute magnitudes plotted against different sorts of "color indices," of which there are almost too many different kinds. Theoreticians prefer a plot of actual luminosity against actual surface temperature (which relates strongly to spectral class), where the stellar luminosities are expressed in solar luminosities or in absolute visual magnitudes corrected for invisible ultraviolet or infrared. We can reasonably well transform one kind of diagram to another, and all the forms are ultimately redundant.

However it is plotted, dwarfs, giants, supergiants, and many other kinds of stars are quickly recognizable from subtle hints in their spectra, in part because of density differences. A giant may have the absorptions of a particular element or molecule enhanced or diminished relative to the same absorptions found in the spectra of dwarfs of the same temperature class. Astronomers have expended great effort in documenting such variances. Once we know the kind of star (class A dwarf, M supergiant, etc.), the HR diagram gives its absolute magnitude, which when coupled with apparent magnitude (through the magnitude equation) provides distance. We can tell to fair accuracy how far away any star is; we can use stellar spectra to trek the whole Galaxy.

whence cometh the classes?

Why are there so many kinds of stars? What underlying principles are involved? Russell and others, not knowing quite what to make of the great differences among stars, came to the logical (but utterly erroneous) conclusion that stars began normal lives at the bright hot top of the main sequence and then cooled and dimmed downward along it. The idea, wrong as it was, had a lasting effect. Hot stars of any class came to be called "early" and cool ones "late." As if we were hosting a real star party (not one in which amateur astronomers gather to view the sky), class A1 Sirius is an inconsiderate early dwarf, while class M1 Betelgeuse is a late red supergiant who presumably can sit anyplace he wants. The vocabulary makes the stars seem to "come alive," much as does the sky on a sparkling clear night.

Though Russell initially thought that the line of dwarfs represented an age sequence, masses derived from binaries show that the main sequence is in fact strictly a mass sequence. O stars have masses upward of 10 times solar. From there the main sequence descends through the A stars, with masses about double that of the Sun; down through the K dwarfs, with masses just below the Sun's; and on to the M dwarfs, which are all below one-half solar. Since the main-sequence Sun runs on hydrogen fusion, by analogy so do the rest of the main-sequence stars.

The low limit for stars that, like the Sun, operate on the proton-proton chain lies in the overlap between the end of class M and the beginning of the new class L, at 0.08 solar mass. Below that limit (which includes much or most of L and all of T), internal temperatures are too low to allow full proton-proton fusion to run, though the more massive of them can still fuse their natural deuterium. Our star party now also hosts substars, or *brown dwarfs* (never mind that they are actually deep red).

From class M through G, the luminosity (L) climbs roughly as mass (M) cubed, $L(suns) = M(suns)^3$. Earlier (hotter) than class G, the exponent jumps to 4, $L(suns) = M(suns)^4$. A star ten times as massive as the Sun is 10,000 times brighter. This *mass–luminosity relation* allows us to extend the main sequence to 120 Suns for the earliest O3 dwarfs. Above 120 solar masses or so (the number is hardly secure), the odds of stellar creation are probably nearly non-existent; and if such beasts did form, the outward pressure exerted by their great luminosities (by the force of light itself) would very quickly tear them apart. *Mass*: that is what the main sequence is all about.

Giants and supergiants do not fit the main-sequence's mass-luminosity relation. Compared to the dwarfs, they are too bright for their masses. Something else must be afoot. To help find what it is, first look at the giants' polar opposites, at stars that are terribly small and faint. When we finally look at the extended family, we see that it is indeed age.

white dwarfs

As Sirius, the "Dog Star" of the constellation Canis Major, traces its relentless path against the more distant background stars, it also wobbles back and forth over a period of 50 years. Procyon, in Canis Minor, behaves the same way, but with a different period. The only thing that could possibly make a star deviate from a straight-line path (really, just a tiny section of a huge curved Galactic orbit) is gravity. Mutually orbiting bodies revolve about a common center. Sirius and Procyon must have companions that continually deflect them. Sirius's mate, discovered in 1862, appears as a faint eighth-magnitude dot next to its overwhelmingly bright neighbor, the ten magnitudes of difference (a brightness ratio of 10,000) making it devilishly difficult to see. Procyon's companion fell to observation shortly thereafter. Orbital calculation showed Sirius's companion, Sirius B, to have a mass close to that of the Sun, while the naked-eye star, Sirius A, weighed about double solar. (Do not confuse double-star vocabulary with the spectral classes. In a double, the brighter is called A, the fainter B; in a triple the next down is C, and so on. Sirius A is coincidentally a class A star but one thing has nothing to do with the other.)

These discoveries, while certainly interesting, were not particularly disturbing. Compared against the main sequence's mass-luminosity relation, giants are too bright for their masses; in contrast these new-found stars were too dim for theirs. A faint star with a mass of the Sun could simply be one that had cooled off, its life nearly over. The mystery deepened, however, with the discovery of another such body orbiting the fourth magnitude star 40 Eridani. This one was far enough from its companion for its spectrum and color to be observed. It was white, a class A star. It was hot. To be hot and dim at the same time, it had to be small, very small, not much bigger than Earth. What to call such a star? "Dwarf" had been used. This star—and as proved later, the companions to Sirius and Procyon—were more or less white, so they and myriad others eventually discovered were called *white dwarfs* to discriminate them from the ordinary dwarfs of the main sequence. Sirius A is a dwarf that is white. Sirius B is a white dwarf. They are nothing alike.

White dwarfs are indeed dying, cooling stars, and like other kinds of stars, span the HR diagram across a great range of temperature—and therefore color. Only the first ones discovered were actually "white" or at least seemed to be. At 27,000 degrees Kelvin, Sirius B is hot enough to be blue-white. Some are hotter yet, while others down near 4000 degrees Kelvin take on cooler colors. Thus we see blue white dwarfs, white white dwarfs, and orange white dwarfs, illustrating the problem of naming before fully understanding.

But naming, however bizarre, is a minor issue compared to the physical problem raised by the discovery that white dwarfs had the same temperature range as ordinary dwarfs. For Sirius B to hold a solar mass and yet be the size of Earth, it must be incredibly dense. Density is defined as the amount of matter per unit volume. The density of water is 1 gram per cubic centimeter. (In fact, the gram is defined as the mass contained in 1 cubic centimeter of water.) The

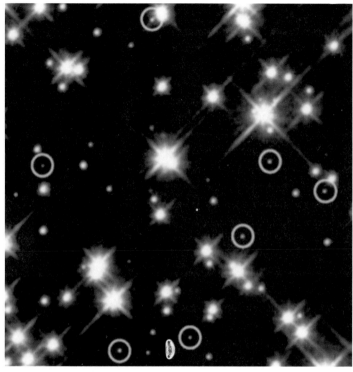

Tiny white dwarfs lie amongst the much
brighter, though quite ordinary, stars of the
globular cluster Messier 4. Only the
Hubble Telescope can see them within
such a dense cluster. (H. Richer, University
of British Columbia; StScI; and NASA.)

average density of Earth (the mass of the Earth divided by its volume) is 5.5 grams per cubic centimeter. But that of surface rocks is around 3 grams per cubic centimeter, implying that there must be something very heavy inside the Earth to compensate. The most common heavy element is iron. A third the mass of Earth lies in an iron core (most of it liquid) that extends halfway to the surface, its reality supported by the existence of the Earth's magnetic field and by analysis of earthquake waves.

The Sun's average density is about that of water, increasing from less than a thousandth the density of the Earth's atmosphere at the solar surface to an astonishing 140 grams per cubic centimeter at the center. This enormous density, over 10 times that of lead, is produced by the weight of the overlying layers and is required to run thermonuclear fusion. Since the Sun's radius is about 100 times Earth's, the solar volume is a million times greater. So, if you create a white dwarf by squeezing the Sun into the Earth, the average density goes up a million times, to a million grams—a metric ton—per cubic centimeter. And it must be greater at the white dwarf's center. At that density you could stuff a car into a cube only 1.5 centimeters—6/10 inch—on a side, and the Earth into a ball only 230 kilometers (140 miles) in diameter.

How? Easy. Atoms are almost all empty space, the electrons 100,000 times farther from the nucleus than the nucleus is across. Remove the electrons (that is, ionize the gas, the natural state of most matter within a star) and let gravity squeeze. Nuclear density, achieved when the nuclei (mostly protons) "touch" one another, is 10^{15} grams per cubic centimeter, a billion times greater than white-dwarf density: Earth jammed into a sports stadium. Surprisingly, the question is not how can white dwarfs be so dense. It is why are they not much *denser*. What holds them up? If full thermonuclear fusion were running inside, they could not be so small. What then keeps gravity from squeezing the star practically out of existence? The answer, coming from a variety of directions, provided powerful support for advancing physical theories.

There are not "sciences," but "science." The distinctions among them, their divisions at universities into academic departments, while probably necessary administratively, is artificial, as each supports and overlaps the others. The study of modern astronomy requires atomic and nuclear physics, and knowledge of gas behavior, of fluid flow, of chemical behavior. Physics, to a good extent born of astronomy, continues to be confronted with astronomical observations that it must explain. Even biology and astronomy meet in the search for life in the Universe and answers to the question of how life developed here. The explanation of white dwarfs lies in the "new physics" of quantum mechanics, now a century old, which holds (among other precepts) that at the most fundamental level, energy and mass come in discrete packets, or *quanta*.

Electrons, protons, and other "particles" are not little balls of matter, though in some instances they do behave that way. In other instances, they behave like waves, very much as do photons, the particles of light. In fact, they are all waves and particles at the same time, neither one nor the other, but both, combined in a way intuitively unfathomable but mathematically describable. Since a particle is also a wave, and the wave can have no boundary or edge (there is no "shore line" in the Universe), the particle cannot be pinned down. We can say only where it is most *likely* to be. But it may not be there. It may in fact be anywhere.

This concept provides the actual key to understanding thermonuclear fusion. Even under the extreme conditions at the solar core—15 million degrees Kelvin and 140 grams per cubic centimeter—two hardball protons cannot be slammed close enough to enable the atomic forces to make deuterium. But the particles *can* find themselves so close that their wave properties take over, so close that the particles have a high probability of "jumping along their waves" (there really is no human analogy) to sufficient proximity to allow the nuclear reaction to run.

Particles of light, photons, carry particular amounts of energy that strictly relate to their wavelengths or wave frequencies (frequency the number of wave crests passing an observer per second). Photon energy is equal to frequency times Planck's constant, h (named after Max Planck, a German physicist who was instrumental in developing quantum theory in the early twentieth century). In units of grams, centimeters, and seconds, h is a mere 10^{-27}. Individual photons do not carry much energy: a 100-watt light bulb produces about 10^{20} photons—a hundred million trillion—per second. Planck's constant is fundamental to the Heisenberg Uncertainty Principle (after Werner Heisenberg, another German pioneer of quantum theory), that the uncertainty in the position of a particle times the uncertainty in its momentum (speed times mass) is approximately equal to h. If you know the speed exactly, you have no idea at all of where the particle is; and if you know position, the speed could be anything.

Particles have a variety of properties: mass, charge, even spin, a kind of rotation that, like energy, is quantized to specific units. The Pauli Exclusion Principle (named for the Austrian physicist Wolfgang Pauli), related to all these other concepts, shows that within a box with a volume roughly equal to Planck's constant cubed, you cannot have two identical particles. (The box has six sides, three mutually perpendicular sides in real space, three more in "momentum space," where momentum is mass times velocity: don't even think about trying to visualize it.) You can have two electrons of the same speed within the box, but they must spin in opposite directions. You can pack more electrons into a spatial volume, but they must enter at ever higher momenta, or speeds. White-dwarf matter is ionized, the atoms stripped of their electrons. When the density reaches the point at which two electrons of the same momentum are packed into the box, it is said to be "degenerate," and you can pack no more at that particular speed. The outward pressure exerted by these degenerate electrons fights off gravity and keeps the star from shrinking. White dwarfs are therefore degenerate. Even orange and blue ones.

census

Astronomy, like other sciences and perhaps more than most, is afflicted with the severe problem of "observational selection." Nature shows us what she wants. Elephants are quite obvious, microbes not. So from simple observation we might make the terribly wrong assumption that there are more elephants than microbes. The naked-eye sky is populated with stars of the bright upper main sequence, and with giants. Of the 40 brightest stars, over half are giant or dwarf A and B stars, five are orange K giants, and two are red supergiants. There are no class M dwarfs. In fact, in the naked-eye sky of 8000 stars there are no M dwarfs at all. To the eye alone, hot bright stars and cool giants dominate the Galaxy.

Reality is exactly the opposite. Our naked-eye view does not take in a full sample of a specific volume of space. Even with sophisticated equipment, such a sample is quite difficult to obtain. The problem lies in the enormous range of stellar luminosities, from top to bottom a factor over a trillion. Supergiants, giants, and bright A and B dwarfs can be seen over immense distances, whereas class M (and L) dwarfs are so dim that they would have to be right next door to be observed without a telescope, and none is. The brightest M dwarf, only 13 light years away, is still only magnitude 6.7, just under the usual limit for unaided human vision.

If we take a small volume around the Sun over which we can observe and count the red dwarfs we find they completely dominate. Confining ourselves to the classic sequence (O through M), of the 50 stars within 15 light years, 36 are M dwarfs! There is but one A dwarf, Sirius, 8.6 light years away. There are no B stars and neither giants nor supergiants. There are, however, three white dwarfs (showing that these are reasonably common, at least far more so than giants) and a growing count of L and T brown dwarfs. The closest B star is 24 light years away, the nearest O star 800 light years! M and L dwarfs far outnumber any other kind of actual star, and the brown dwarfs may be more numerous yet. Class A and B stars and giants of any class are clearly quite scarce, and the O stars are very few and far between. We simply see the luminous stars over great distances, making them seem far more populous than they really are.

Of all stars from class O to M, 70 percent are M dwarfs. Stars like the Sun make up around 10 percent. B stars constitute a mere 0.1 percent. Of all stars, only 0.0001 percent are of class O! They are exceedingly rare. Nature seems to love the small, and does not at all like to make hot massive stars. But nature has also been kind. As we will see, we owe our very existence to the rare O stars. Yet as in so much of life, there can be too much of a good thing. Too *many* O stars would make life impossible. There seem to be just enough.

With these intriguing thoughts hanging in the air, look now not at how the stars have been responsible for life in general, but more personally at their significance to our own lives in the sense of human culture and history. The wonder of the O stars, as well as the revelation of the role played by stellar aging, will wait until later.

nightly parade

Watch the stars and see the Earth rotate. Stars rise above the eastern horizon while others set in the west. Note the position of a star now, then again in 20 minutes, and see how far it has moved along a daily path parallel to the celestial equator. If you are in the northern hemisphere, face north and find the north celestial pole (marked by the modest star Polaris) up from the horizon by an angle equal to your latitude. If in the southern hemisphere, face south to the south celestial pole. Directly above the Earth's stationary north pole, the north celestial pole is stationary too, the stars continuously wheeling around it, some never reaching the horizon, making terrestrial rotation obvious.

Watch the stars and see the Earth revolve around the Sun. Over the year, the Earth orbits through 360 degrees in 365 days, just under 1 degree per day, causing the Sun to skim along its ecliptic path at the same rate. Each night at a given time we face in a slightly different direction, 1 degree around the circle. Note the positions of the stars tonight. Come back in a week at the same time and they will have shifted 7 degrees to the west. Over the course of the seasons, the stars opposite the Sun march across the nighttime sky.

Moving daily, moving yearly, the stars provide us with natural clocks and calendars, ways of telling when to do the important things of life, like planting, harvesting, and having good times. Positioned in the sky from east to west, north to south, some lying on the equator, others near the pole, the stars give us not just time but direction. Though there are now technological layers between the stars and ourselves, they still serve as guideposts of time and space.

Formed into patterns, into the familiar constellations, they remind us of days gone by, commemorate our heroes, help us tell our fables. We relate many of the same stories today, nothing really new, just recast, our nightly narrator the book or the television instead of the fire-lit storyteller. The stars, along with the constellations, assumed names, and became old and valued friends. Pause now in the exploration of the natures of the stars to see their meaning to humanity past and present as we watch their parade.

populating the sky

With some exceptions, stars are strewn at random across the sky, and naturally make patterns. Human cultures everywhere have wondered about the stars and the meanings of these formations. After all, the gods live in the heavens, so the stars must have something to do with the gods and therefore with our fates. (They do of course, but nothing the ancients could possibly have dreamed.) Different cultures used the natural patterns to make different constellation

It is obvious from the paired bright stars (Castor on top, Pollux on the bottom), and the symmetry of the long rectangle stretching to the right, how the zodiac's Gemini, the Twins, received its name. Not all constellations lookas much like what they are supposed to be. (Akira Fujii.)

figures, and told different stories about them. The constellations seen and named in China were nothing like those of the Incas of Peru or the natives of North America. Our "western" constellations were born many thousands of years ago in the ancient lands of the Middle East. From there, the constellations passed to the hands of the early Greeks, where they took on Greek lore. Beautifully codified in poetry and prose, from Homer on through classic and later times, they were passed down to us, 48 ancient constellations whose origins remain largely unknown.

In ancient societies, the constellations had deep mystical meanings. Though today many professional astronomers cannot recognize constellations—and even disdain them—they are important. The starry sky, as anyone who views it from a dark mountaintop will attest, is huge. The constellations partition the celestial vault into smaller units, allowing stars and many other kinds of astronomical objects to be recognized and named. Equally important, the constellations draw us out to look, much as they did our ancestors, their beauty and charm encouraging us to appreciate the stars for what they really are, and to look deeper into their true meanings. There in the Dipper are stars like the Sun; there in Boötes, orange Arcturus, a dying giant; there in Orion, Betelgeuse, a red supergiant that may someday explode.

By far the most important of the ancient 48 are the 12 that lie along the ecliptic, the apparent path of the Sun. These are the figures of the zodiac, all but one living ("zo-diac," from the same root as "zo-ology"). Hear them call out from ancient times, many of them profound and powerful fertility symbols: Pisces (the Fishes), Aries (the Ram), Taurus (the Bull), Gemini (the warrior Twins), Cancer (the Crab), Leo (the Lion), Virgo (the Virgin), Libra (the Scales, or Balance), Scorpius (the Scorpion), Sagittarius (the centaur Archer), Capricornus (the Sea Goat—you have to know the story), and

Aquarius (the Water Bearer). And the odd one, Libra, at one time the outstretched claws of the Scorpion, *used* to be a living thing. There are 12 because the Moon goes through its phases roughly 12 times a year, so each successive full, or new, Moon is in the next constellation over.

The position of the Sun within the zodiacal constellations told us when to plow and reap. Some 2500 years ago, the vernal equinox was in virile Aries, the autumnal equinox in Libra's balance pans. These positions still echo in newspaper astrology columns, in which Aries leads the list. (The equinoxes have since moved westward one constellation as a result of a long wobble in the Earth's axis, a fuller story told below.) The zodiac gave names to Earth: in ancient times, the summer solstice was in Cancer, the winter one in Capricornus, hence the Tropics of Cancer and Capricorn, the latitudes at which the Sun passes overhead on the first day of northern summer and winter. Set within the figures are great jewels: the orange giants Aldebaran of Taurus and Pollux of Gemini, the red supergiant Antares of Scorpius, the blue-white main-sequence dwarfs Regulus of Leo and Spica of Virgo, and Gemini's sextuple white main-sequence Castor.

Beyond the zodiac, the greatest fame goes to Orion and his crew and to the Bears of the north country. Orion, straddling the celestial equator, is seen everywhere, his figure commonly considered human. The Greeks named him after their great hunter, the Arabs after a now-mysterious woman. First-magnitude supergiants, red Betelgeuse and blue-white Rigel, mark Orion's shoulder and foot, three dazzling blue-white stars (the Arabs' "String of Pearls") his belt. Followed across the sky by his faithful hunting dogs, Canis Major (containing Sirius) and Canis Minor (with Procyon), he stands upon his prey, dim Lepus, the Hare. He was placed in the sky by the gods when Diana was tricked by her brother Apollo into killing him. In another story he was stung by Scorpius, so the gods placed them opposite and out of sight of each other.

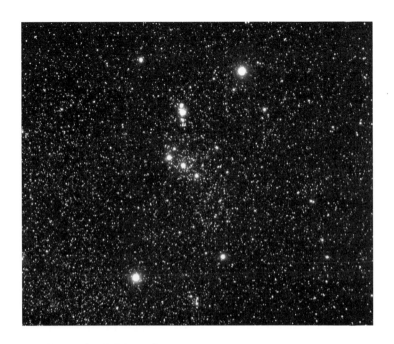

One of the most famed of all constellations,
great Orion, the Hunter, clearly reminds one
of a human. One of the brightest figures in
the sky, it has two first magnitude stars,
Betelgeuse at upper left and Rigel at lower
right. In between is the Hunter's three-star
belt, from which dangles his sword.
(Akira Fujii.)

Arching high in northern climes is that most beloved of figures, the Big Dipper (in Britain, the Plow). Not part of the ancient 48, the Dipper is an informal constellation—an "asterism"—that is part of Ursa Major, the Greater Bear. Its front bowl stars (one an orange giant, go look) point at Polaris at the end of the handle of the faint Little Dipper, an asterism in Ursa Minor, the Smaller Bear. The Dippers' handles form the Bears' tails. How do you put a bear into the sky? Grab its tail and whirl and throw with all your might, stretching its short natural tail into a long one.

Surrounding the Hunter and Bears are dozens of other figures: Andromeda, lovely daughter of vain Cassiopeia the Queen; her rescuer Perseus; and Cetus the Sea Monster bent on devouring the young lady. Near the Big Dipper lies Corona Borealis, the Crown of the abandoned princess Ariadne; above Orion Auriga, the Charioteer; above Scorpius, Ophiuchus, the Serpent Bearer, representing the healer Asclepius. There they are, the gifts of the ancients: heroes, birds, and beasts, all to admire and pass down to those who follow you.

filling in the blanks

The ancients named only what they easily saw. East of Orion between the Dogs lies a drab area containing only fourth- to sixth-magnitude stars that presumably waited patiently for the gods to make a new great constellation. This and other "unformed" regions were indeed filled, but in a different way. The new science of astronomy that burgeoned with Copernicus's revelation of the true nature of the Solar System demanded that the spaces be attended to. From around 1600 to 1800, astronomers had great times inventing new constellations from faint stars, carving up old constellations to fit new ideas, and filling in the part of the deep southern hemisphere that could not be seen from classical lands (never minding that the locals had already done that).

As the ancients made constellations from their cultures, so did the "modern" astronomers, providing the sky with a variety of machines and with both real and mythical animals of current interest: the obvious Telescopium and Microscopium, Fornax (the Furnace), Sextans (the Sextant), the eponymous Lynx and Phoenix, Camelopardalis (the Giraffe), Monoceros (the Unicorn, that between the Dogs), and dozens of others. Not all are obscure. Crux, the famed Southern Cross, was taken from lower Centaurus, the ancient Centaur; and south of the Big Dipper lies Coma Berenices (Berenice's Hair), the open cluster that makes an entire constellation.

So many constellations were invented that the whole lot could hardly survive. Do not mourn, for example, the fortunate demise of Machina Electrica (the Printing Office) and Musca Borealis, the Northern Fly (bad enough that Musca Australis, the Southern Fly, still buzzes). Nationalistic figures designed to curry favor—the Oak of Charles II, the Scepter of Frederick the Great—were hardly popular with the other sides. All was put in order in the early twentieth century by the International Astronomical Union, which kept 38 modern constellations and drew rectangular boundaries around all of them. With Jason's vessel, huge Argo, broken into Puppis (the Stern), Vela (the Sails), and Carina (the Keel), the sky now holds 88 official constellations, plus the great number of assorted informal asterisms (as well as rejected constellations) that are enough to amuse and edify for a lifetime.

Though famous, Crux, the Southern Cross
(at right), is a modern constellation, one not
visible from ancient northern lands, and
constructed only in recent times. It contains
two first magnitude stars. To the left are the
two first magnitude stars of Centaurus, Alpha
(the closest star to the Earth) at far left, Beta
to the right. (Akira Fujii.)

naming the stars

The human eye alone can see about 8000 stars, and with a telescope far more; professional imaging raises the number to the billions. We must keep track of them somehow, whether for ancient story and fortune-telling, to guide our spacecraft, or to study their physical natures and lives. Therein lies a story of human culture and history.

At first we gave them proper names, much as we would our sons, daughters, and pets, usually names that described something about the stars. Thus that luminary of Canis Major, the brightest star of the sky, was named Sirius by the Greeks, from their word for "searing" or "scorching." Following Ursa Major is Arcturus in Boötes (the Herdsman), meaning the Bear Driver ("arktos" the Greek word for Bear; hence the "arctic"). Latin followed. "Regulus," in Leo, means "the little king"; Spica, in Virgo, "sheaf of wheat," representing the harvest.

Though the Greeks passed the ancient constellations down to us (with Latin names assigned by their Roman conquerors), few star names are actually Greek or Latin. Most are Arabic. In the demise of the Greco-Roman civilization, the Arabians of northern Africa took up the cultural cudgel, translating Greek texts (many of which were eventually lost) into Arabic, thus preserving an immense amount of human learning and wisdom. Though the Arabians had their own constellation lore, they, like the Romans, admired Greek culture, and the Greek figures largely replaced their own. (Theirs still shine by way of numerous asterisms such as "the virgins," four stars that make the southern portion of Canis Major.) The Arabians then named a great number of stars after their positions in the classical constellations.

The writings of the Greeks leaked back to Europe through Arabic translations and were retranslated into Latin, often by people who had no deep knowledge of Arabic. As a result, many of the names were misrepresented, mistranslated, even assigned to the wrong stars. Thus we find a mixture of perfectly good Arabic names and those that sound Arabic (often beginning with the article "el" or "al") but that would quite mystify anyone from modern or ancient Arabic lands. At the top of the comprehensibility scale we find "Deneb," meaning "tail," at the end of Cygnus, the Swan. (This star also serves as the top of the Northern Cross.) The sky abounds with tails: Deneb Algedi in Capricornus, Deneb Kaitos in Cetus, Denebola in Leo. Then there are Betelgeuse and Rigel, which are reductions of phrases that originally meant "the hand" and "foot" of Al Jauza (the "Central One"), the Arabs' version of Orion. These names make some sense, but by themselves are unintelligible. At the bottom are stars like Albireo (at Cygnus's head), whose names were hopelessly mangled from other terms and made to sound Arabic.

Hundreds of proper names can be devilishly difficult to recall; a more logical scheme was needed. The good work was supplied by Johannes Bayer around 1600. Bayer, who created the *Uranometria*, the greatest pre-modern star atlas, applied lowercase Greek letters to stars within a constellation, chiefly in order of brightness, attaching the possessive form of the constellation's Latin name. Capella, the brightest star of Auriga, thus became Alpha of Auriga, or Alpha Aurigae (possession implied by the ending); the second brightest of Cetus, Deneb Kaitos, became Beta Ceti, and so on. There are many exceptions to the brightness rule, as Bayer broke some constellations into parts, and named stars in order of progression in others. The Dipper stars are lettered front-to-back in spite of magnitude. Other orderings defy logic; the Beta star of Canis Major, Mirzam, is not close to being second brightest, that honor belonging to Adhara, which received Epsilon.

Nonetheless, Bayer's system survives to this day, with first-magnitude stars commonly going by their proper names, fainter ones by their Greek letters. Until, of course, we drop below the last letter, Omega. Bayer extended his system with lowercase, then uppercase Roman letters, but these were superseded by the astronomer John Flamsteed and his numbers. In the early 1700s, Flamsteed was hired by the English Crown to create an accurate star map to be used for navigation. He listed the stars within a constellation from west to east according to their right ascensions (their positions east of the vernal equinox). The stars were later numbered, possibly by Isaac Newton and Edmund Halley (who to Flamsteed's fury seem to have pilfered an early version of his catalogue). Flamsteed numbers are commonly used after the Greek letters run out, hence the name of the first parallax star, number 61 of Cygnus, or 61 Cygni.

In addition to all of these designations, astronomers use a variety of catalogue numbers divorced from the constellations, usually progressing according to right ascension. At some dim point, coordinates alone suffice. All in all, many millions of stars have some kind of name. We know them and know where they are.

your fortune please

The Planetary System is closely (though hardly perfectly) confined to a plane that is close to the Sun's rotational equatorial plane. From Earth, the planets move along their orbits near the ecliptic, and almost always within the constellations of the zodiac. No one in ancient times knew what the planets were. The life-giver, the Sun, surely either a god or run by a god, followed the zodiac. What then about the other moving bodies? Might they not be, or represent, gods too? What about their places of residence, the mysterious constellation figures through which they moved? Perhaps if our forebears knew and could interpret the placement of the gods within

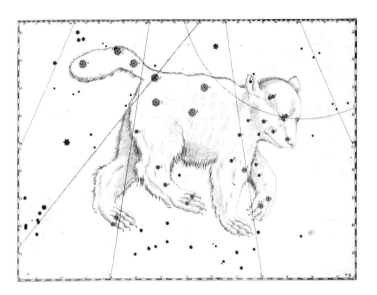

Ursa Major, the Greater Bear, walks beneath
the pole in this representation from the most
famous of old star maps, the *Uranometria* of
Johannes Bayer, published in 1603. The
brighter stars are named with lower-case
Greek letters. (Rare Book Room and Special
Collections Library, University of Illinois.)

them, and relative to one another, they could learn the gods' dispositions and know something of their own futures.

Questions like these gave rise to astrology, astronomy's "evil twin," which purports to use the stars to tell of human affairs. The two grew up together, astrology a strong motive in astronomers' attempts to learn the natures of the planets' motions and their placements relative to us and to the Sun. Although some of the ancients knew the difference between astronomy and astrology, the two really began to diverge seriously only when we started to understand the true nature of the layout of the Solar System and yearned to know the physical properties of the Universe.

Whether astronomers like it or not, astrology truly is a powerful influence in human affairs, as so many believe in it, even run their lives by it. Where possible it needs to be addressed in astronomy books. In its simplest modern form, your personality and fate are determined by the location of the Sun at the time of your birth, not within an astronomical constellation of the zodiac, but within an astrological "sign." The constellations are a motley lot, some big, some small; the Sun spends 44 days in Virgo, a mere five in Scorpius, moreover taking 17 days to traverse the modern boundaries of the constellation Ophiuchus, which cut across the ecliptic. The signs are uniform 30-degree segments that overlaid—and did not quite fit— the constellations some 2500 years ago. The signs took their "magical" properties from the constellation names: Taurans are stubborn and bull-headed, Capricorns stolid, Leos brave, Geminis duplicitous (they never tell you the bad parts).

The signs are locked to the equinoxes and solstices. The Earth, however, has an equatorial bulge caused by its rotation. Lunar and solar gravity pull on the bulge and cause the Earth's tilted axis to wobble, or *precess*, around the orbital perpendicular over a 26,000-year period. The wobble rotates the celestial equator, and causes the points of intersection with the ecliptic, the equinoxes (solstices, too), to move westward. As a result, in the time since their inception, the astrological signs have been carried westward by one full constellation, the sign of Capricorn now overlapping Sagittarius, that of Cancer on top of Gemini (hence the Tropic of Cancer, though the solstice is in Gemini). Somehow the movement does not affect the signs' meanings.

It is obvious that not all people born under Leo are the same. Allowance for individual differences requires a scheme that is vastly more complicated and uses all the planets and their positions. Such "real" astrology is much harder to put in newspapers and magazines than the simplistic "sun sign" astrology, which is disdained by serious fortune tellers. The planets all have qualities that fit the gods they represent: Venus is beautiful, Mars contentious and warlike, Jupiter bountiful. Each sign has properties that affect the behavior of the planet-as-god in which it resides. The planets also influence each other, the effect of Mars on human affairs different when it is close to Jupiter rather than close to Saturn.

The influences of the planets must somehow extend to the person. From the Earth, a dozen "houses" surround you, placed in the sky along the ecliptic. Beginning at the east point of the horizon, they plunge below the Earth, climb back up above the west point, and cross the sky to the east, six above the horizon, six below. Each house controls some human quality: one love, another money, and so forth. Your fortune is determined by the placement of the planets within the signs, by the angles they make with one another (and the Sun), and by where they fall within the houses at the time of birth. The whole picture is called your "native," or birth, "horoscope." Moreover, what you should do at any moment is controlled by the planets' current locations. The casting of a horoscope thus becomes complicated and expensive, requiring the exact time and longitude of your birth. Not surprisingly, the whole scheme is so complex that it is open to a variety of interpretations, so the astrologer is able to find some configuration that will indicate what you want to hear.

Though to true believers a confrontation with logic will have little effect, it could make a difference for those on the fringe, or who just never thought about the subject a great deal. Constellations are not real objects, nor are they permanent. They are chance patterns caused by alignments of stars usually lying at very different distances from Earth. They change and disintegrate with time as a result of the stars' movements around the Galaxy, and appear different from different stars. The magical properties of the signs were assumed at a seemingly special time in human history, as precession moved the signs past the constellations. The signs' magical properties are in the mind of the beholder, not in the stars.

Even granting that the stars and planets could directly affect us, how could the influences be conveyed? Gravity transmits energy and information across empty space. A common argument in favor of astrology is that the Moon and Sun pull on, and move, the Earth's waters, causing tides. Since the human body is largely water, the Moon must also be able to influence us personally, as then must all the planets. But tides are well understood, not magical, and they work only because the Earth and its oceans are so large that there is a significant stretching force. Even in a large lake there are no tides; and people are so small that tidal effects are nonexistent. What then about radiation? Jupiter is a powerful source of radio emissions, one of the strongest in the sky. Other planets radiate as well, and the Sun pours it down on us. But even with all this radiant power, the lights, radio stations, and electrical wiring around us present a far stronger radiation bath, and all are quite benign.

At this point, a believer in this ancient mystical lore may simply invoke unknown forces, the ever-popular "cosmic vibes." But even against these there is a defense. Rather than ignore the subject, scientists have explored statistical relations between horoscopes, personalities, and fates. To no surprise, the experiments turn up null results. There is no mystical relation between stars and people. The real relation, that we owe our existence to them in ways that can be understood and explored, is vastly more beautiful.

the timekeeper

We tell time daily by keeping track of the Earth's axial rotation, and yearly by following Earth's orbital revolution around the Sun. To learn the time of day and year, we use outside references: the Sun and stars. During the day, they follow apparent daily paths that make them rise in the east and set in the west. During the year, the Sun plies the ecliptic against the background zodiacal stars. (In spite of not seeing the stars in the daytime, we know exactly where they are.) Since we cannot see the Sun and stars all the time, we substitute clocks and calendars, both synchronized with the apparent solar and stellar movements.

The celestial meridian is a circle in the sky that passes through the celestial poles and your zenith, the point over your head. Its intersection with the horizon defines the directions north and south, and it thereby divides the sky into eastern and western hemispheres. As the Earth rotates, everything in the sky must cross the meridian both above and below the visible celestial pole. As the Sun and stars cross above the pole, they reach their highest elevations above the horizon; unless they are sufficiently close to the visible pole, most stars make their lower passage out of sight, after setting.

Divided for ancient reasons into 24 hours, the "solar day" is defined as the time it takes the Sun to go from one crossing of the

Stars swing around the sky's south pole in
response to the Earth's rotation. The
positions of the stars in the sky give both
the time of day and the location on Earth.
(© Gregory G. Dimijian 1993.)

meridian and back through 360 degrees. The Sun crosses the meridian below the horizon at the starting point at midnight, or 0 hours, and again above the horizon at 12 hours noon. For every 15 degrees the Sun moves on its path past the meridian, the time advances by one hour, the clock a model for the moving Sun. (Clocks move "clockwise" because that is the direction in which the Sun appears to move.) If you use a 12-hour clock, between midnight and noon the times are *ante meridiem* (A.M.), Latin for "before noon", and between noon and midnight *post meridiem* (P.M.). All we need to set the clock is to measure the westward angle that the Sun makes with the meridian, or easier, to start the clock at noon, when the Sun is highest in the sky.

We might learn all this in elementary school. Later come the unsettling revelations. The Sun is a poor timekeeper. As a result of the slight eccentricity—oblateness—of the Earth's orbit, which causes slight changes in our distance from the Sun, the Earth moves at a variable orbital speed. It goes fastest when it is closest to the Sun (in January), and slowest when farthest (in July). (The 3.5 percent difference has no practical effect on the seasons.) The Sun therefore must move at a variable rate along the ecliptic. This phenomenon, plus the tilt of the axis, which causes the Sun to move not just east to west, but north to south and back, results in a day (and hour, minute, and second) of variable length. To keep pace, you would have to continuously advance or set back your watch. To circumvent the problem, astronomers invented a fictitious *mean sun* that rides the celestial equator and keeps an average pace with the real Sun. The mean solar time on our clocks can be as much as a quarter of an hour off that told by the real Sun, one of the reasons sundials, which tell time by solar shadows, rarely seem to tell "the right time."

Moreover, we live on a spherical body. For the moment, ignore the Earth's rotation. As you move east, the Sun must appear to move west relative to the celestial meridian. If the Sun moves west, the time gets later. The change is 1 hour for every 15 degrees of longitude (360/24 = 15) and 4 minutes for every degree; at the Earth's equator, 1 second of time corresponds to 460 meters. The people who live in the next street west have a different time!

When travel was slow, who cared? But with the advent of fast train travel, each station had a different time, creating a mess for schedulers. So, since there are 15 degrees in an hour, why not divide the Earth longitudinally into 15-degree-wide belts within which everybody keeps the same *standard* time? When we move from one belt—time zone—to the next, we change the hour but not the minute. The system works beautifully, and is another reason for "bad" sundials, which tell the time locally. A worldwide "Universal Time" (though it is unlikely that Arcturians use it) is finally defined as the mean solar time at Greenwich, where longitude begins.

The day can also be defined by the successive passages of a *star* across the meridian. The Sun is constantly moving to the east along the ecliptic past the stars at the rate of not quite a degree a day. Because a degree corresponds to 4 minutes of time, a star's passage around the sky (from the celestial meridian and back) will take about 4 minutes less than will the solar passage. The day according to the stars (the *sidereal* day, from the Latin word for "star") is therefore about 4 minutes (more accurately 3 minutes 56.56 seconds) shorter than the solar day. The difference is caused by the Earth going around the Sun. For the same reason, we see the stars shift west by about 1 degree per night, 30 degrees per month, when observed at the same time as told by the Sun, giving us the seasonal parade of the constellations. Because the Earth is in motion around the Sun, the sidereal day is the *true* day that would be measured by an outside observer.

We can also tell time by the stars by dividing the sidereal day into 24 hourly units. But since the sidereal day is shorter than the solar, so sidereal hours are shorter than solar hours. Because stars actually move relative to each other, it is best to use a point that is defined among all of them, the vernal equinox. *Sidereal time* is defined by its position relative to the celestial meridian in the same way that solar time is defined by the Sun. A solar clock tells us where the Sun is in the sky. A sidereal clock tells us where the vernal equinox is, and because we know the positions of stars relative to the equinox (through their right ascensions and declinations), the sidereal clock tells us where to find the stars. No observatory can run without a sidereal clock; one is built into the computer that runs the telescope. Because we always know where the Sun is relative to the equinox (through the date), sidereal and solar time can be converted into each other. And because the Sun is a bright extended source whose position is hard to measure, sidereal time, which is readily converted to solar time, is actually the world standard. The stars, not the Sun, tell us when to get up in the morning.

The Sun and stars have long given humanity the ability to count longer intervals. Every 29.5 days, the Moon disappears from the sky and begins its phase cycle with the thinnest of crescents in the west, the classic new moon. Thus were ancient astronomers able to keep track of the months. Every 29.5 days the Moon stands opposite the Sun, allowing us to see its full illuminated face, the full moon. The months are commonly named after them: the grass moon, thunder moon, harvest moon. In older times, the elders would stand under the sky to call out the first sighting and the beginning of the moon, or "month." Once the length of the cycles were understood, people could simply count the days and construct calendars for when the Moon was not seen. Today we stretch the months a bit to fit the year.

Years could be counted by when the stars appeared in the sky. The rising of the Nile was told in ancient times when Sirius, the Dog Star, was first visible in the morning hours as the Sun, always moving easterly, moved away from it. The ancient Greeks knew when to plant by the Seven Sisters' first morning rising. Even today, the cycle known, we count the days. After 365 of them, the Sun returns about—but not quite—to its starting position to end or begin the year. The Earth's daily rotation, however, told by the stars and Sun, is quite unrelated to its annual revolution, the number of days not dividing evenly into the year. The Sun takes 365.24219... days to traverse the ecliptic from the vernal equinox and back. From Caesar's time, the common calendar adds an extra day every four years, making the average 365.25 days; and by order of Pope Gregory XIII in the late sixteenth century, it drops three 366-day leap years every 400 years (in century years not divisible by 400), making for a near-perfect average of 365.2425 days. (The English-speaking world did not convert to the Gregorian calendar until the eighteenth century.)

Atomic clocks can keep track of time to accuracies of 1 second in over a million years, and the Earth's orbit is known to immense precision. Do we still need the stars? We do if we wish to set the clock to fit human endeavors. Tides raised by the Moon slow the Earth, making the day gradually longer. Other effects—shifting of ice caps, winds on mountains—affect the rotation in unpredictable fashions. We must therefore keep setting the common clock; and how should we do it other than with the positions of the stars or related celestial objects? Always, always we return to the stars.

the navigator

Now you know when. Do you know where? You could look on a map. But who made the map, and how? And what do you do if at sea, out of sight of land? Look up. There are the guiding stars, in daylight the Sun, our own star. Directions on Earth are related entirely to the stars. But you could use a magnetic compass to find north. Yes, if you want to go toward the magnetic pole in northern Canada. To go to *true* north, face the Earth's north rotation pole. How, though, when the rotation pole is over the horizon? As an advertising balloon might lead you to the used car dealer below, so the north celestial pole, flying high above the terrestrial pole (as seen from the northern terrestrial hemisphere) shows you where it is.

Assuming you are in the northern hemisphere, find the point of zero rotation, the north celestial pole, closely indicated by Polaris, the North Star, at the end of the Little Dipper's handle. Face the north celestial pole and you face north, the direction perfectly defined by the intersection of the celestial meridian and the horizon. To your back is south, to your right east, and your left west, where the celestial equator hits the horizon. Walk toward the north celestial pole, and when it stands directly overhead you are at the Earth's north rotation pole. The elevation of the pole above the horizon always equals the observer's latitude. Whether in the northern or southern hemisphere, keep the visible celestial pole at the same elevation and you can move precisely east and west, maintaining constant latitude. Early navigators knew these rules well and roamed the world with them.

But they roamed with danger. They could travel east or west, but knew little about their exact east-west positions—that is, their longitudes—which measure displacements relative to Greenwich or any other accepted location. The rescuer is time. As you move east, the Sun and stars appear to move west, and the local time becomes later; as you move west, time gets earlier; the advance and retreat of the clock (ignoring rotation for the moment) is continuous with longitude. Time differences (with hours and minutes converted to degrees and their subdivisions) exactly equal longitude differences. If you know your local time by the stars as well as the time at Greenwich (the Universal Time), you know your longitude! Once navigators had good clocks that could keep accurate Greenwich time during long and tumultuous sea voyages, available from around 1770 on, they could finally travel the Earth freely, locate land masses, and on land make maps that meant none of us would ever have to get lost again.

Look up at night to see the stars wheeling overhead. Their positions in the sky depend on the time of day and year, and upon where we are on Earth. Pick out three of the brighter ones. There is only one place on Earth where those three will be in those positions relative to the horizon at that moment. Knowing the Greenwich time thus gives you your position. Watch the Navigator stand on the deck and measure the elevation of three stars and then calculate the ship's latitude and longitude; on land watch the Surveyor do the same.

Yes, we now have satellites—the Global Positioning System (GPS)—that with timed pulses allows a radio receiver to triangulate exactly where we are, right down to the meter. The system can only improve. No longer must the Navigator feel the cold spray of ocean, no longer must the Navigator and Surveyor wait for clear skies. But once they did, and they got us to where we are today. And to know where the satellites are in a modern world, to keep the beat of the clock set to the Earth's rotation, we must still track the stars. Always, always we return to the stars.

chapter 5

stellar surprise

Giants, supergiants, dwarfs, white dwarfs—all populate our sky. In spite of their imposing characteristics, the friendly vocabulary eventually makes these stars seem quite normal. With the eye alone, we look upward to find examples of all the brighter classes making our constellations; indeed, we go so far as to give these stars familiar names. And they look serenely down upon us.

Or do they? We have yet to examine the sky close up, to see what these stars might be doing both in front of us and behind our backs. Moreover, we have yet to explore what might lurk between the standard kinds of stars, bodies that—like the modern constellations—"fill in the blanks" between the obvious and expected. So enter now a different world, one of surprising strangeness and, at times, of awesome violence. Explore now stars that differ mightily in chemistry from the norm; stars that vary in brightness, some changing over a huge range, others varying with amazing speed; stars that throw matter back and forth between them; stars that make O dwarfs look cold and white dwarfs seem gigantic. Here we will meet exploding stars as well as others that simply disappear from space.

Odder yet than any of these is the final realization that without such bizarre behavior, we could not, would not, exist.

the alchemist's dream

The Alchemist tried to turn lead to gold, his efforts in part leading to modern chemistry. He would be disappointed to know that even the most advanced chemists still cannot perform the act and never will. Instead, to his amazement, he would find that chemical elements are transformed not by chemists, but by stars, and that the more likely route would be from gold to lead. Not only all our gold, but with the exception of a few light ones, *all* our chemical elements came from the stars.

And we can prove it by observation linked with theories that neatly explain what we see. The most obvious examples of stellar transmutation were discovered in the mid-nineteenth century (though astronomers could hardly know it at the time). Father Angelo Secchi, the Italian Jesuit astronomer whose early classification scheme laid the foundation for the modern spectral sequence, OBAFGKM(LT), found that red stars divide into two distinct kinds that exhibit different spectral absorptions. The more common, our modern M stars, display complex absorption lines of oxides, particularly of titanium and vanadium. However, in the spectra of rarer deep red stars, these metallic oxides vanish and are replaced by carbon compounds. Called class N by Harvard's Pickering, they track the temperatures of the cool M stars all the way down to 2000 K or under.

A warmer version was later assigned to class R. They are now known simply as *carbon stars*.

The classic carbon stars are giants. Modern analysis shows that M stars have about twice as much oxygen as carbon, so there is plenty to make prominent metallic oxides. In the carbon (N) stars the ratio is reversed and then some. Since carbon has a powerful craving for oxygen, no other oxides can form, and the leftover carbon combines with itself. In between there is another cool breed that also tracks the M stars, class S, in which carbon about equals oxygen. Here, a spoonful of leftover oxygen mates with zirconium, which it likes better than titanium. The result is strong zirconium oxide absorption. Carbon and S stars violate the basic principle that stars have common compositions. Since stars are born from clouds of gas in which oxygen dominates, something must be going on to transform the surface compositions, some combination of processes that became clear only with the development of theories of stellar aging.

Absolute proof of elemental transformation was achieved with the discovery of the element technetium in giants of all three cool classes (M, S, and N). Technetium, element 43 (43 protons), has no stable isotope. The longest-lived one lasts only a couple of million years, far shorter than any of these stars live. There is no natural technetium on Earth. (What little we do have is manufactured for medical purposes as a tracer for abnormal cells.) And there should be none in stars. But there it is. It can only be made there. The unusually large zirconium abundances seen in S stars suggests that this element, and likely others, are also made in stars. The only known route is through nuclear reactions whose byproducts are brought from the hot stellar innards to the visible surfaces.

In contrast, the stars of the Galaxy's halo are severely depleted in all heavier elements. Astronomers, simple souls, commonly lump all the elements into three portions: hydrogen, helium, and "metals," a catch-all term for "everything else," lithium through uranium. The metal abundance of a typical globular cluster is only about a hundredth that found in the Sun. At the extreme, there are single stars in the halo that are a hundred times less metallic: they have almost no metals at all. How did nature go from such stars to the Sun? Some process must *make* metals. We know that stars manufacture carbon and many other things. Earlier generations of stars must have made the metals inside themselves and given them to the Sun— and to us.

Of course, nature enjoys fooling us. A host of "chemically peculiar" stars, most of them warmer class B, A, and F, display numerous and truly weird abundances that include depletions of some elements and huge enrichments of others such as neodymium and europium. Among the most extreme are the mercury-manganese stars, in which the abundance of mercury can be as much as 100,000 times solar. These strange abundance patterns are not caused by any kind of nuclear alchemy, but by physical separation, something nobody expected a star to be able to do. The Sun's outer layers are constantly stirred by violent convection. In warmer stars, convection disappears, and the stellar atmospheres, the outer layers where the spectral absorptions take place, become calm. Some heavier atoms sink, pulled down by gravity; others efficiently catch outgoing radiation and sail upward.

The separation of elements reaches an extreme among the already odd white dwarfs. In most we see nothing but absorptions from hydrogen. The gravitational pull is so extreme that heavier helium, which should constitute 10 percent of the gas, is pulled completely out, leaving nothing but light stuff. In other white dwarfs, however, we see *only* helium! Given that the helium should settle, these must not have any hydrogen at all. The astronomer's difficult task is to distinguish the effects of these separation processes from those that actually make heavier atoms through nuclear reactions and bring them to the stellar surfaces. The result of all the research is that we can see the chemical elements created almost before our eyes.

variations on a theme (some by hidin')

You do not have to look hard to find stars that *do* something. Second magnitude Algol, Beta Persei, drops in brightness by a magnitude every 2.87 days. But the star deceives. It does not really vary, but only seems to, as it is an eclipsing double in which one star hides in back of another during its orbit. Fourth magnitude Delta Cephei, however, is different. Though it too periodically varies by about a magnitude (over 5.37 days), it is distinctly single; it changes all by itself. Single, but hardly singular, it is the prototype for thousands of known *Cepheid variables* found both in our Galaxy and in others, their variation periods running from 1 to 100 days. The classic ones are all class F and G bright giants and supergiants, massive stars in the upper middle of the HR diagram. If you plot their HR positions, they define a tight strip of obvious instability. Several can be followed with the naked eye, including Eta Aquilae (the brightest of them) and Zeta Geminorum. Even Polaris is a Cepheid, though with a brightness range too small to be seen by eye.

Cepheids are the beating hearts of the cosmos: they pulsate. If you could watch one from nearby, you would see it change its radius, surface temperature, color, and luminosity all at the same time. Such a star cannot find a place for itself, and alternately expands and contracts. The change in radius leads to a change in temperature, and both conspire to change the luminosity. A Cepheid is not unlike a child on a swing. Left alone, the child would dissipate his or her swinging energy and slow down. To maintain swinging regularity, the youngster—and a Cepheid—must be pushed. The star is not really changing the luminosity coming from its core, which knows nothing of the effect; the pulsation takes place only in the outer layers. Hydrogen and helium at the surface of the star are neutral; but inside, the temperature climbs to the point at which these two elements become ionized, electrons becoming stripped from atoms. The helium ionization layer, where the element is partially ionized some 100,000 kilometers below the surface, acts as a natural valve. As the star pulsates, the layer changes its transparency and its ability to absorb radiation, which in turn causes a cycle of expansion and contraction that keeps feeding back on itself. To become a Cepheid, the layer must be at the right depth, and the HR diagram's instability strip is where it happens. Oddly, some stars in the strip are *not* Cepheids. We have no idea why.

Cepheids are important in telling us something of stellar structure. But their real value goes back to 1912 and Henrietta Leavitt of Harvard. Our Galaxy's two small companions, the Magellanic Clouds, are of immense importance to astronomers, as the member stars of each are equally far away (about 150,000 light years), allowing stellar features to be compared independent of distance. Leavitt quickly noted that the Clouds' brighter Cepheids took longer to pulsate, no surprise really, as the brighter stars must also be the larger ones, and larger ones take a longer time to make the cycle.

The real surprise is how tight and linear is the correlation. On a graph of average absolute visual magnitude versus the logarithm of the period in days, the stars line up along a nice straight line. We therefore need only measure a variable's period to find its absolute magnitude and hence an accurate distance. The problem is calibration. The Magellanic Cloud Cepheids give us *apparent* magnitude against period. To scale the graph properly, we need either to find the absolute magnitudes of some Cepheids in our Galaxy by knowing their distances or find the distances of the Magellanic Clouds. Both have been done. Some Cepheids are within parallax range. Others lie in our Galaxy's open clusters, whose distances can be found from comparison of their HR diagrams with clusters whose distances (and absolute magnitudes) are known from parallax. We can do the same kind of comparison with the stars of the Magellanic Clouds.

Why care about the distance to a given Cepheid? Cepheids, as supergiants, are so luminous they can easily be seen in other galaxies far beyond the Magellanic Clouds. Though hard to find and study because of the crowds of stars, they are worth the effort because once you have one, you have the galaxy's distance. Such stars allowed Edwin Hubble to measure the distance to our neighbor, the Andromeda galaxy, in 1924, thus helping establish as fact the expansion of the Universe. The discovery with the Hubble Space Telescope of more than 800 Cepheids in galaxies as far as 100 million light years away has led to an accurate estimate of the rate of expansion and an age for the Universe of 14 billion years or less.

Cepheids range from around 25 times the size of the Sun to as big as the orbit of the Earth. Though red giants are, as a class, visually less luminous than Cepheids, their coolness gives the brighter of them even larger dimensions. Bright, cool, red giants also share the Cepheids' instabilities and have similar internal drivers. Because of their sizes, they take much longer to pulsate. The first one found—indeed the first variable found, in 1574—so amazed those who watched it that they called it Mira (as in "miracle"), the Wonderful Star. Buried in the neck of Cetus the Whale (as Omicron Ceti), Mira—half again the size of the Martian orbit—varies so much it disappears from naked-eye view, changing from around third magnitude to tenth (far below human vision) and back over a period that averages 334 days.

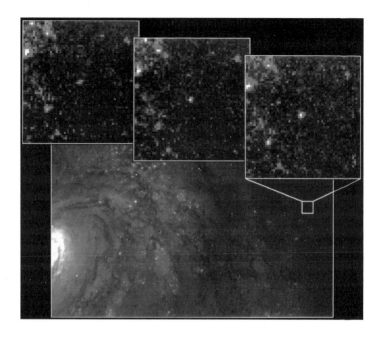

Within the spiral galaxy M 100 (below) lie
Cepheid variable stars (one seen above), whose
variations can be studied with the Hubble Space
Telescope. Establishment of their periods give
their true luminosities, which, when compared
with the viewed brightnesses, gives distances,
showing that M 100 is 56 million light years
away. (W. L. Freedman, Observatories of the
Carnegie Institution of Washington; STScI;
and NASA.)

Mira is the leader of tens of thousands of other "Mira variables" that dot the sky. It is not that they are so common in the halls of stellar life, but that they are so luminous they too can be seen for immense distances—Mira itself lies 420 light years away. Period variations run from 50 or so days to an amazing 1000. As expected from the Cepheids, they follow a period-luminosity relation (though one not so tight), the longest-period stars shining with 100,000 solar luminosities. Though like most red giants they are of class M, they can also be carbon stars or class S, like Chi Cygni, which roams from almost fourth magnitude to a dim fourteenth. In line with their huge sizes, Miras are also the coolest of the red giants, indeed among the coolest of all stars; their only rivals are the dimmest stars at the absolute bottom of the main sequence. Ranging to spectral class M10, temperatures fall to 2000 degrees Kelvin. As a result, absorption bands from molecules dominate their red light.

Mira variations are deceptive. They vividly show how limited we are with just our "optical" eyes, and underscore why astronomers must examine nature at all wavelengths. Miras really only change their total luminosities by about two magnitudes, not all that much more than Cepheids. A great part of their radiation is in the infrared, where we cannot see it. Most of the dramatic drop in apparent visual brightness is caused by a relatively small temperature decrease that sends a disproportionate amount of radiation from the visible spectrum to the infrared, and in addition causes a thickening of the molecular blanket that helps block whatever visible light remains.

Also generally invisible to the human eye is a Mira's most important characteristic: its powerful wind. Though the solar wind blows at a mere tenth of a trillionth of a solar mass per year, it causes notable effects within the Solar System, buffeting gaseous comet tails away from the Sun and creating the auroras. Imagine what life on Earth would be like were the wind to increase 100 millionfold. Life might be impossible, as the wind would strip the planets of much of their atmospheres.

All bodies have an *escape velocity*, the speed a projectile would have to have in order not to come back. Increasing the mass of a body increases the gravitational field and consequently the speed needed to leave. An increase in radius, however, increases the average distance of the projectile from the body's atoms and therefore *decreases* the local gravity and the escape velocity. The Miras are so huge that their escape velocities are very low, so it does not take much to get matter off their surfaces. Theory says that their pulsations set up sonic-boom-like shock waves (their effects seen in spectra) that drive mass upward, where in the chill of space some of the molecule-rich gas condenses to dust. Comets, which shed both gas and dust under the action of sunlight, reveal what happens next. The solar wind pushes a comet's gas tail backward. Light, however, also exerts pressure. It is the pressure of sunlight acting on solid grains that produces the long swooping dust tail that so captivates the eye. The immense luminosity of a Mira variable shoves its dust outward. Friction with the gas moves it along too, and the whole dirty mess shuffles off at 10 to 20 kilometers per second. The dusty wind can be so thick at the extreme that the star hides itself, peeking through only at longer wavelengths, in the invisible infrared.

Here is a nexus in astronomy, where we begin to see the linkages that exist among stars, and between stars and the Galaxy. Stars change only slowly. A star can be in the Mira state for so long that it loses much of itself to space, eventually almost the entire envelope that surrounds its nuclear-fusing furnace. A good fraction of the gas of interstellar space consists of recycled stars. The fleeing dust takes on a life of its own, the little particles—so small you could not see them even in the palm of your hand—acting as seeds onto which interstellar atoms and molecules can condense. Look to the Milky Way some clear night and admire the dark lanes and bays that are not absences of stars but patches of the Milky Way obscured by dust clouds, vast clumps of dirty gas that block the distant view. Most of the dust came from Mira variables. New stars condense from these clouds. But they cannot unless their birth clouds are cold. The dust blocks the light from heating starlight and provides the refrigeration. Miras are mothers to the stars.

Huge numbers of different kinds of other variables populate the heavens. The instability strip plunges downward through the entire HR diagram. The lowered metal content of halo stars changes the transparency of the gas and makes halo giants warmer. Where the strip encounters these smaller, warmer giants, especially among those that crowd globular clusters, we find the *RR Lyrae stars,* which oscillate in less than a day. All about absolute visual magnitude zero, they too provide excellent distance indicators.

Stellar instabilities are suppressed, but not eliminated, where the strip approaches the main sequence near class A, so even here stars can chatter lightly away. Down below, among the hot white dwarfs, the stellar pulsation driver still takes up the whip. Even these dense stars can be made to vibrate, but because of their tiny sizes they have periods of only minutes. Moreover, they do not pulsate uniformly, but in sections, some parts of the stars moving outward while others move inward. Similar odd variations are seen among hot stars of class B that are just above the main sequence.

The variety to be found is extraordinary! Every few years, a warm, luminous naked-eye class F giant in the constellation Corona Borealis suddenly disappears from view as it plummets unpredictably to as deep as fourteenth magnitude. Its near-vanishing act comes from an erratic wind that produces immense amounts of carbon "soot" that eventually dissipates, allowing us again to see the star beneath. At the top of the HR diagram, supergiants are so large that most become unstable to some degree. These wanderings are so erratic that no one is able quite to predict what such stars will do. Betelgeuse, Orion's great M supergiant, varies with several periods, from days to six years.

Some stars change over decades in concert with immense mass ejections that shame those of the "ordinary" giants. Eta Carinae, shining 5 million times more brightly than our Sun, reached the minus first magnitude in 1848 even though it is 8000 light years away. By 1880 this 100 solar mass star—or stars (it may be double)—had fallen to below naked eye visibility after ejecting more than an entire Sun's-worth of dusty gas into space.

pitch and catch

The most predictable of all these variable stars are still the hiding, eclipsing doubles. To be an eclipser, the components of a double must be close together, or the odds of one passing in front of the other would be vanishingly small. When two stars are close enough, they can raise tides in, and distort, each other. In extreme cases, the smaller will cause the larger to take on a tear-drop–shaped surface at which the gravity is effectively zero, the point of the teardrop aimed toward the other star. Matter can then be thrown from the larger star through the teardrop's point to the smaller star, the catcher gaining notable mass at the expense of the pitcher.

In Algol, the bright and classic eclipser, the incoming matter rams directly onto the receiver star, light from the flowing stream adding irregularity to the double's variations. In other pairs, the incoming mass flows first into a surrounding disk, from which the receiver star accretes it. This *accretion disk* is also bright enough to

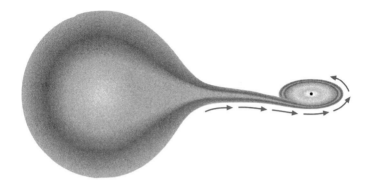

Double stars whose components lie close to
each other can exchange matter, sometimes
substantial amounts of it. Here, the gravity of
a dense compact star raises great tides in a
larger star of comparable mass, stretching it
to the point where the combined gravities
cancel, and matter can flow from the pitcher
to the catcher.

contribute to the system's total light, and can be quite unstable, leading to unpredictable variations.

An extreme involves a white dwarf in tight orbit about an otherwise ordinary dwarf. Even the pure hydrogen atmospheres that encompass most white dwarfs are not very deep. Matter flowing from the main-sequence dwarf onto the white dwarf thickens and heats the layer. When it becomes dense and hot enough, the layer erupts in a thermonuclear explosion, making the double star hundreds of thousands of times brighter than the Sun. From Earth we see a "new star," or *nova,* erupt into the sky. About once a human generation one will be close enough to hit or exceed first-magnitude brightness. After some years, the exploded debris can be seen expanding around the system at speeds of a thousand kilometers per second. Yet the white dwarf is so dense that its innards remain unaffected. It merely relieves itself of its overburden; and then, after 100,000 or so years the star does it all over again. Observation supports, indeed leads to, such a scenario: ex-novae are all close doubles.

Vast numbers of combinations and possibilities present themselves. At the limit, one star can completely destroy the other. Stellar siblings, like their human counterparts, do not always get along.

smoke rings

Confusing enough that we have white dwarfs that are not really dwarfs, dwarfs that are white but are not white dwarfs, and such oddities as orange white dwarfs. We also have *planetary nebulae* (*nebula* is Latin for "cloud") that have little or nothing to do with planets. The name may be a misnomer, but an understandable one, as their discoverer, William Herschel (who in 1781 discovered the planet Uranus), meant only that their shapes reminded him of the planets. Through a small telescope, they appear as small, nearly round, almost featureless disks; through a large one they exhibit surreal complexity.

Planetary nebulae are soft clouds of light that surround hot blue stars. Thousands are known, most within the confines of the Milky Way, but they can be seen thronging other galaxies as well. They range in size from nearly stellar to monsters that appear nearly as big in the sky as the full Moon, physically from a fraction of a light year across to sizes that span the distances between neighboring stars. They are among the most popular sights at observatory viewing nights, teachers and amateurs alike showing the famed Ring Nebula in Lyra, the Dumbbell in Vulpecula, and many others. The stars in the middle are almost the hottest known, with temperatures that *start* at those of O stars, around 25,000 degrees Kelvin, and climb to over 200,000 degrees Kelvin. At such temperatures, they emit most of their light in the invisible ultraviolet. If that is taken into account, they are among the Galaxy's most luminous stars, shining thousands of times more brightly than the Sun.

The ultraviolet light from such a star is so energetic that it strips the atoms in its surrounding nebula of their electrons, ionizing the gas. When the atoms recapture the electrons, much of the energy is released in the visual part of the spectrum, causing the nebula to glow. Because they release energy instead of absorbing it, the glowing gases of planetary nebulae display narrow emissions—*emission lines*—at the same wavelengths at which stars show absorption lines. Analysis of the relative strengths of the emissions reveals gas temperatures around 10,000 degrees Kelvin and typical densities of a few thousand atoms per cubic centimeter. Though the nebulae are vacuous—indeed are better vacuums than can be created in terrestrial labs—their large sizes allow them to include a good fraction of a solar mass and to shine quite brightly. Most important, the emission lines

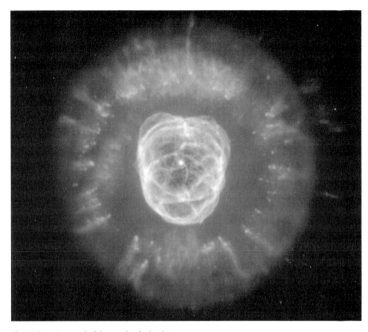

The "Eskimo," a wonderful example of a bright
planetary nebula, is made of matter ejected from,
and illuminated by, the hot dying star in the
center. The extraordinarily complex Hubble image
shows ejected gases ramming into each other.
(A. Fruchter and the ERO Team, StScI, and NASA.)

allow the determination of chemical compositions. Most of the nebulae have compositions much like that of the Sun, but here and there are some that are enormously rich in certain elements, particularly helium, nitrogen, carbon, and undoubtedly heavier elements not directly observable.

The nebulae are expanding into space at around 20 kilometers per second, similar to the speeds of winds around Mira variables. At that rate, they live for only 50,000 or so years, a mere snapshot in time compared to star lives. They must therefore be created at a great rate to produce the number we see. Also, their central stars have about the same luminosities as the Miras. Some nebulae, and their stars, too, are carbon-rich. The Miras, some of which are carbon stars, are losing the outer envelopes that embrace their nuclear-burning hearts. Could it be that the nebulae are the results? The planetary nebulae are quite logically stripped Miras, the central stars the ancient stellar cores. We see stellar aging almost before our eyes.

Planetary nebulae are among the most beautiful objects the sky has to offer. Detailed imaging shows shells within shells, intricate whorls, loops, and wisps of colored gases, and opposing "bullets" of matter that somehow have been shot from the central stars. As the likely ejecta of Miras, they reveal the impressive complexity of the winds. They also show us how at least some matter is deposited back into interstellar space for the creation of new stars. And they also show directly how fresh elements find their way into the cosmos, ultimately to be given to stars like the Sun.

explosion

There are other celestial depositors, and some are awesome. In 1572, Tycho Brahe, who made the observations of the planets that led to Kepler's laws of planetary motion, witnessed the details of the first "nova" to be recorded in Europe. It shone in Cassiopeia with the brightness of Venus and appeared in the daytime sky. Thirty-two years later, Kepler recorded another. Historical research on old Chinese records unearthed more of these brilliant interlopers, one in 1006 so bright that it illuminated the night. On the average of every 200 years, one of these stars, far brighter than common novae, lights the sky.

An odd nebula, Number 1 in Charles' Messier's famed eighteenth century catalogue of "non-stellar objects," lies off the horns of Taurus. It falls exactly where the Chinese recorded their brilliant "Guest Star" of 1054, which had shone with the light of Tycho's Star. Its sprays of gaseous tentacles gave it the name *the Crab*. Doppler measures of the Crab Nebula showed it to be expanding at a rate of 10,000 kilometers per second, far faster than the shell of an ordinary nova. Once distances could be known, particularly for such erupting stars observed in other galaxies, these exploders were seen to be far more luminous than novae, beyond a billion times solar, not novae but *supernovae*. There is no continuity in luminosity or in anything else between novae and supernovae. Supernovae must therefore be something quite different. Here there is no mere surface detonation. The brightness dictates that practically the entire star must go. They are among the most destructive events known in nature—and at the same time among the most creative, as they help generate the compression of the interstellar gases that creates new stars.

The star that exploded near the center of the
Whirlpool Galaxy (M51) in 1994 could easily be
seen, even though 20 million light years away.
At its peak, the star was as bright as 100 million
Suns, and rivalled the light of a small galaxy.
(R. Kirshner, Harvard University; STScI; and NASA.)

Spectra show that supernovae, as well as their debris, are highly enriched in heavy elements, including iron. Given that we can see only a fraction of the Galaxy's supernovae because of thick interstellar dust, there must be three of four events per century, the rate going far back into time. And given the amount of iron each one produced, they are all we need; all the iron in the Universe, as well as most of the other chemical elements, seem to have come from them.

The rarity of observed supernovae—none has been seen in our Galaxy since Kepler's Star—means that they must be studied in other galaxies, where they occur in abundance. Astronomers long ago recognized two kinds of these exploding stars. Some—Type I—contained no hydrogen, while others—Type II—did. Type II explosions take place exclusively in the disks of galaxies. Type I are found anywhere, including the galaxies' halos; moreover, they occur in galaxies that have no disks at all. What distinguishes a galaxy's disk? Look into the Milky Way, at Orion and Scorpius, made of brilliant blue, massive O and B stars. None are found in the halo or in non-disk galaxies. Type IIs seem to be the progeny of massive stars, an observation now well substantiated by theory. Most, if not all, O stars blow up. Give thanks for the O stars' rarity, as a supernova within a few light years could do serious damage to Earth.

Supernovae of Type I cannot be made by O and B stars (technically Type Ia, as there is a Ib that is indeed associated with massive stars). Ordinary stars like the Sun cannot explode, so doubles are invoked to explain Type Ia supernovae. But only compact stars have enough gravitational force to generate such energy. Given their numbers, white dwarfs seem somehow responsible. Whatever their cause, Type Ia events all hit about the same absolute visual magnitudes, around –19 (more than a billion solar luminosities) as calibrated from parent galaxies whose distances are known from Cepheids. They are so bright they can be seen at distances that exceed a billion light years, providing distance indicators that can reach far into the Universe, allowing us to probe its nature. Type IIs on the other hand have allowed us to probe some of the Universe's most amazing "citizens."

as small as they get

In the 1930s, the perspicacious astronomer Fritz Zwicky suggested that the only way (it was not, but he was close) to blow up a star was to use gravity to collapse the core of a larger star into a much smaller one. But it had to become *so* small—a few tens of kilometers across—to provide the needed energy, that it could only be made of atomic particles crushed together into neutrons. The Crab Nebula, lying in the Milky Way, is spattered with stars. One close to the center stands out. It has no absorption lines in its spectrum. Could it be Zwicky's neutron star? Discovery had to wait for a long time.

Revelation came, as it so often does, from an entirely unexpected quarter. Observations with radio telescopes in the late 1960s accidentally revealed celestial objects that were producing short, sharp, steady bursts of radiation with periods between the bursts of only a few seconds. The beats were incredibly regular, as good as the best clocks. The radiating objects, called *pulsars*, had to be some kind of star, as nothing else could be so visible. But big stars like Cepheids pulsate—change their radii and brightness—over days. Some white dwarfs pulsate with periods of minutes, but they are still far too large to pulsate in seconds. Moreover, ordinary stellar pulsations cause a star merely to change brightness; the stars are always visible. In between the very fast radio bursts from the pulsars, however, there was nothing; they simply disappeared from view. Moreover, ordinary

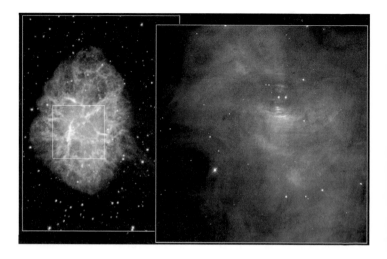

The Crab Nebula, easily visible off Taurus's
southern horn, is the exploded remnant of
the Chinese "Guest Star" of 1054. In the past
thousand years, the Crab has expanded to a
diameter of 12 light years! At its center lies
the dense remains of the star itself, a spinning
neutron star the size of a small town that
clearly and violently stirs the surrounding
gases. (Left: Palomar Observatory, Caltech;
right: J. Hester and P. Scowan, Arizona State
University, and StScI and NASA.)

pulsation could not be as deadly regular as the radio bursts turned out to be.

Rotation to the rescue. Not much can change the rotation of a stellar mass over a short time, over mere days. It alone provides the requisite precision. But a body as "large" as a white dwarf cannot possibly spin with a period of a few seconds either. To do so, the pulsar had to be only a few kilometers across... it had to be a *spinning* neutron star, the term pulsar yet another misnomer. The densities of such stars must be enormous, nearly nuclear density itself, 10^{14} or so grams—100 million tons—per cubic centimeter. Against them white dwarfs are huge and vacuous. Like white dwarfs, however, neutron stars cannot be held up by nuclear fusion. They too are degenerate, supported not by degenerate electrons but by degenerate neutrons that cannot be forced any closer together.

Along with the great density goes an immensely powerful magnetic field a trillion or more times that of Earth's. The magnetic fields of most bodies are tipped with respect to their rotation axes. The monster field causes radiation to be beamed out from the magnetic poles, which are swinging wildly in space. If the Earth is in the way, we get a short, brief blast, the neutron star now seen as a pulsar. One pulsar was found to "burp" 30 times per second. It was in the middle of the Crab Nebula. Optical telescopes set up to record quick flashes showed it to be the star without absorptions: Zwicky's neutron star at last. Exploding stars must produce pulsars.

A pulsar's magnetic energy source ultimately lies in its rotation, so as it radiates it must slow down. And sure enough, over long intervals, the pulse periods lengthen. A rapidly spinning newborn pulsar has enough energy to radiate across the spectrum, the Crab pulsar producing X-rays, even gamma rays. As pulsars brake, the high energies drop off first, and by the time their periods are measured in seconds, they radiate only low-energy radio waves, like the first ones discovered. When the periods hit more than around five seconds, the little stars drop out of sight. Since 1000 pulsars are known, and since they all decay, dead ones, nonpulsing neutron stars, ought to be everywhere, roughly consistent with the number of supernovae that take place. One quiet neutron star found with Hubble has a temperature of a million Kelvin, making it the hottest star known.

Other phenomena reveal the pulsars' power. Remember the game of pitch and catch? If a quieted, "retired" pulsar has a sufficiently close companion, its mate will be working on the pulsar all the time, feeding matter to it. The stolen mass hits the neutron star to the side with such force that it gradually speeds back up, eventually hitting rotation speeds of hundreds of times per second as it is reborn as a "millisecond pulsar." Such a star would simply be a blur to the human eye. The spin-up stops when the companion evaporates. Pulsars are so dense they have crusty, crystalline surfaces that bulge at the equator. As they spin down, changes in their internal forces require sudden adjustments in their structures, which make the pulsars speed up a bit. At the most extreme (or so it is believed), magnetic fields 1000 trillion times that of Earth can crack the crusts and produce monumental bursts of energy. One burst 20,000 light years away produced so much energy that it ionized the Earth's upper atmosphere and severely agitated orbiting satellites. Similar readjustments may be involved with bursts of gamma rays that pour in from distant reaches of the Universe so far away that the parent galaxies appear only as smudges to the biggest telescopes on Earth. The moral of these tales? Stay very far away from neutron stars.

Yet pulsars, as bizarre as they seem, are not the end of the story. The Earth's escape velocity is 11.2 kilometers per second, the speed needed for interplanetary exploration. In an impossible experiment, squeeze the Earth down. You are then on the average closer to the Earth's attracting atoms, so the escape velocity goes up. The escape velocity of a tiny neutron star is not that far below the speed of light. If a body could be made small enough for the escape velocity actually to reach that of light, light would be trapped, and the body would disappear from view, becoming an invisible *black hole*. Einstein perceived gravity differently, as a distortion in *spacetime* (the four-dimensional structure in which we live), as a "well" in spacetime created by mass. When you fall in a gravitational field, you are rolling down the slope. In relativity, which is the only theory that can actually describe a black hole, the hole would appear as an infinitely deep puncture in the fabric of the Universe, from which nothing can escape.

Do black holes exist? A class B supergiant in Cygnus (Cygnus X-1) that has no business doing so radiates powerful X-rays. Doppler observations show it to be in orbit about something many times more massive than the Sun, more massive than neutron stars are found to be, yet something that cannot be seen. Half a dozen doubles similar to Cygnus X-1 have been found. A black hole itself would be invisible; but once again, in the ultimate game of pitch and catch, matter passing from the supergiant to the black hole would be heated to immense temperatures, high enough to make X-rays, before the infalling mass is lost to the world forever. If exploding stars can make neutron stars, maybe they can make black holes too. Or maybe not. We need to look at the theories.

chapter 6

linkages

It is not enough to know the different kinds of stars. As in life, the journey is as important as the arrival. How are white dwarfs, supernovae, neutron stars, and the like made? What precedes them and what do they become? Can we predict the stages, perhaps point to a star and say what is going to happen to it? Closer to home, can we predict what will happen to the Sun, and when?

With a few exceptions, the stars change too slowly for any effects to be visible in a human lifetime, even in thousands of human lifetimes. Given that no one can watch a star age from start to finish, or even age through any substantial part of its life's pathway, the only recourse is for the observational astronomer to inventory the behavior of what nature presents, then give the data to the theoretician (who might be one and the same). The goal is to make sense of it all through the application of physical principles that have been wrested through earthly studies. We learn how all the different stars came to be by stringing them together with theories of gravity, the atom, electromagnetism, and other forces. And we have been successful. We are learning to know the stars.

the unbounding main

All stars will be, have been, or are main sequence stars, ordinary dwarfs. The main sequence is the HR diagram's broad band of hydrogen fusion and the most stable period in any star's life, where, with some exceptions, not very much happens. And good thing, because without main-sequence stability, real life on Earth would never have had the time to develop and flourish. The luminosity of a star and its position on the main sequence, as well as its life expectancy, depend on its mass. The Sun had enough core hydrogen to last for 10 billion years; 5 billion are left to go. How long will other stars take to begin to die?

The lifetime of a wood fire lengthens as the amount of fuel is increased, but drops as the burning rate goes up (imagine throwing gasoline on it). More formally, the fire's lifetime is proportional to (fuel supply)/(burning rate). In turn, the greater the burning rate, the brighter the fire. Making things as simple as possible, the fire's lifetime is roughly proportional to (fuel supply)/(luminosity).

The same is true for stars. The greater a star's mass, the greater the amount of fusible hydrogen in the core; the stellar fuel supply is therefore proportional to the total stellar mass (M). As a result, main-sequence lifetime (t) is proportional to M divided by the stellar luminosity (L), or t goes as M/L. But from double-star studies, luminosity goes up much faster than the mass (on the average luminosity scaling as $M^{3.5}$). As mass increases, t (now proportional to $M/M^{3.5}$ or $1/M^{2.5}$) must go dramatically *down*. If you double the mass of a star, it will stay on the main sequence only a tenth as long. High-mass stars die first.

We can be far more precise. The main sequence divides neatly at about 1.5 solar masses, near spectral class F2. Cooler than that, stars run on the proton-proton chain, in which helium is built directly from hydrogen. But hotter stars run predominantly on the carbon cycle, in which carbon aids in the fusion of four protons into helium. Given a stellar mass, a theory of internal structure (how temperature and density change with radius), and a knowledge of fusion mechanisms, theoreticians can accurately calculate the mass of the hydrogen-fusing core and the time it will take to burn away to helium. At the high-mass end, dwarf-life will be over in a mere 2 million years, while at the bottom, lifetimes are counted in trillions of years.

The aging effect is obvious when comparing the stars of clusters, and allows the calculation of cluster ages. Some clusters have fully intact main sequences, including hosts of high-mass O stars. They must be very young. The Pleiades is missing its O stars, and the Hyades main sequence starts at class A. From the time it takes their main sequences to burn down, the Pleiades must be around 100 million years old, the Hyades closer to a billion. The oldest open clusters are about 10 billion years old. Because they occupy the Galaxy's disk, that should also be the disk's age. The globular clusters are more extreme, their main sequences having burned away to about 0.8 solar mass, which requires between 12 and 14 billion years. Their home, the Galaxy's halo, must therefore be older than the disk. Since nothing is older anywhere, that too is the age of the Galaxy. That the old globular clusters, as well as halo stars in general, have low metal abundances, fits well with observations that stars make the chemical elements and float or blast them back into space. Younger stars contain the metals produced by older generations, as the whole Galaxy evolves chemically with time.

The clusters show that the Galaxy's halo must have formed first out of the original matter created by the Big Bang. As the Galaxy aged, it gravitationally collapsed to form its current younger disk. Add in collisions and mergers, and you have the messy Galaxy of today. Since galaxies should not have taken long to form after the Big Bang, the ages of the oldest globular clusters should also be about the same as the age of the Universe itself. This age fits beautifully with that derived from the expansion rate of the Universe. Such agreement can be no accident. We are doing something right!

The reliable dwarf stars divide themselves in other ways. If it takes the age of the Galaxy to remove a star of 0.8 solar mass from main-sequence life, then no lighter star (one cooler than G8) has ever died. There is no way to test the aging theories for such stars (except to wait a very long time). At around 10 solar masses, however, there is a sea change. A heavy star's innards develop quite differently from those of the lighter ones. Below around 10 solar masses, stars die relatively quietly; above that limit, they explode. Wherever they fall, however, all stars have profound effects upon us and upon the Galaxy as a whole, even if through nothing but their massed gravity.

roaring giants

Look first to the middle of the main sequence, to stars like the Sun. Dwarfs of all kinds (including those on the high-mass main sequence) are amazing, though perverse, self-regulating devices. They are kept from collapsing by heat generated by internal nuclear reactions. Intuition would say that as the hydrogen supply diminishes, the core should cool and the star should dim. Instead, the lowered numbers of individual atoms in the core (four of H to one of He) cause it slowly to contract, the released gravitational energy making it *hotter* inside. The higher temperature makes the reactions run faster and causes the core to eat slightly into the surrounding hydrogen envelope, both compensating for the increasing fuel deficit. In fact, the effects overcompensate, so a dwarf actually becomes brighter! The Sun should have been 30 percent fainter when it was born, and will be twice as bright as now when the interior hydrogen is gone. One result of this slow aging is that the main sequence is a broad band rather than a line. The other, more personal, effect is that the Earth will be turned into an oven long before main-sequence life is over.

Fly to the future 5 billion years from now and witness that crucial event, when the Sun's core finally becomes pure helium. The central nuclear fire now shuts down, and for the next 1 to 2 billion years the core begins a much faster collapse. Gravity turns up the thermostat even higher, and hydrogen fusion spreads outward into a shell around the now-defunct core. The interior adjustments first cause the Sun merely to expand, forcing it to cool at the surface to below 5000 Kelvin into class K. Then, even though core reactions have ceased, the Sun will brighten by a factor of 1000 or more, blistering the Earth. (If you think that is scary, pity poor Mercury, as continued expansion takes the Sun right through its orbit and perhaps even to that of Venus.) The Sun is now a glowering class M red giant, its central core condensed to the size of Earth and to immense temperature and density.

Nuclear fusion rescues the star. Atoms can fuse themselves to iron with the release of energy. Heavy atoms can break apart—fission —*down* to iron with energy release (causing radium to glow). But you can do nothing with iron; it sits there like a lump, explaining why there is so much iron in the Universe. Progressively higher tempera-

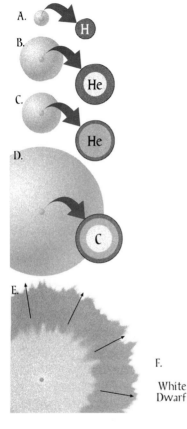

Lower-mass stars have a well-known fate. *(Note: Nothing is to scale in this illustration.)*

A: A sunlike main sequence star quietly burns hydrogen to helium in its core.

B: The central hydrogen has run out, and the dead helium core shrinks within a hydrogen-burning shell, while the outer part of the star expands to make a red giant.

C: Now the helium, still in its hydrogen-burning nest, fires up to burn to carbon and expands while the outer part of the star shrinks some to become a warmer orange giant.

D: The central helium runs out and the carbon core, now within alternately burning hydrogen and helium shells, again shrinks, while the stellar envelope re-expands, eventually becoming unstable enough to pulsate as a Mira variable.

E: The Mira becomes so large and loose that it loses almost all of its outer envelope, nearly exposing the hot core, which eventually lights up the fleeing gases to produce a planetary nebula.

F: When all is over and the planetary nebula has dissipated into the interstellar gases, only a tiny white dwarf the size of Earth remains.

tures are required to fuse heavier atoms. When the core hits 100 million degrees Kelvin, helium fuses to carbon. That is no easy task. Two helium nuclei ram together to form the amazingly unstable isotope beryllium-8 (four protons, four neutrons). Within a hundred-quadrillionth of a second, before it falls apart, the beryllium must pick up another helium nucleus, which results in stable and common carbon-12. Before we knew what they were, helium nuclei were called "alpha particles," so the process is named the *triple-alpha reaction*. A slam with another "alpha" makes oxygen from the carbon. The new energy source stops the core's contraction. Still perverse, our future Sun contracts and dims some. Only a hundred or so times its current luminosity, it will become an orange class K giant, one similar to those that abound in the evening sky.

Most of the energy released in fusing atoms from hydrogen to iron goes out in the first step, hydrogen to helium. Helium to carbon releases much less. Since it has the same job, that of supporting the star, core helium fusion lasts nowhere near as long as hydrogen fusion. After well under a billion years, the helium runs out, leaving behind a dead core of carbon and oxygen. With no internal support, gravity again squeezes it down.

As the core temperature climbs, helium fusion now moves outward into a shell that lies between the carbon-oxygen core and the older hydrogen-fusing shell, and the star repeats its earlier behavior with a vengeance. With the two shells turning their fusion on and off in violent sequence, the star re-climbs into the realm of the bright red giants, becoming even larger, redder, and more luminous than before. The Sun will quite likely reach the orbit of the Earth as its surface temperature descends to 2000 degrees Kelvin. A "second-ascent" star's huge proportions give it an internal structure that makes it unstable. It begins to pulsate, expanding and contracting over a period of a year or so, changing its brightness as it goes. And look!

A new Mira variable dots the sky, spewing gases from its surface in a slow dusty wind.

Though the outlines of the process are the same from the lower limit of 0.8 solar masses up to around 10 solar, the details differ. Stars above around 5 solar masses, which start off as much hotter class B stars and go through the steps much faster, do not brighten so much as greatly cool; these higher-mass stars loop back and forth across the HR diagram, cooling, heating, then cooling. As the more massive and brighter stars enter the HR diagram's instability strip around class G, they may become Cepheid variables. A star might be a Cepheid at one time in its life and a Mira at a different time.

Hydrogen fusion in the outer shell, via the carbon cycle, creates new nitrogen, while helium fusion in the inner shell makes new carbon. If the mass and structure are just right, these by-products can be brought to the surface by convection, and the star can change from a class M giant into an S star and then into a carbon star. Other nuclear reactions go wild. Energy production via minor cycles makes an abundance of neutrons. Having no electric charge, neutrons easily attach themselves to heavy atoms. Caught one at a time, they steadily increase the neutron number, building ever-heavier isotopes until the atom hits one that is unstable and radioactive. A neutron in the nucleus will then spit out a negative electron and turn itself into a proton, thereby increasing the element number. In this way, via *slow neutron capture* (slow as in peaceful, at a leisurely rate), a heavy element can be converted to the one above it, rubidium becoming strontium, then zirconium, niobium, molybdenum, technetium, and beyond. All this stuff can be cycled upward too, into the stars' outer layers, allowing us to see extra zirconium in S stars, and technetium in all sorts of giants before it decays. All is wafted away in the great Mira wind. Most of the carbon in the Universe, as well as large fractions of many other elements, have come to us from the Miras, mothers to the stars and in a broad sense parents to ourselves.

the end: part I

Now what? The carbon-oxygen core is shrinking and heating. From past experience we might expect these elements to begin to fuse to something else. But no: we have neglected mass loss. It is so strong that nearly the entire overburden of inert hydrogen that lies atop the fusing shells is lost. A star like the Sun will lose over 40 percent of itself back to space; an 8 solar mass star will lose over 80 percent. The stellar interior can become no hotter, and the carbon and oxygen are stuck. Almost exposed to the chill of space, the remaining core shuts down its massive cool wind. As a small hot body, it begins instead to blow a much hotter and faster wind (since higher speeds are needed to escape the greater surface gravity).

Imagine the old, massive wind. Laden with dust (silicates or carbon, depending on the kind of star), it is expanding smoothly away, minding its own business. Suddenly it is blindsided by the speedier hotter wind that has caught up with it. Slamming into the slow wind, the fast wind shovels up huge quantities of gas and dust into a dense shell whose shape will depend on how the slow wind was structured when it left the star and how the fast wind behaves. The fast wind, and hydrogen fusion from below, expose more and more of the core. The star heats at its surface, transforming itself from class M to K to G to F to A, all the while buried in its obscuring dusty effluvia. From Earth we see the fleeing dust of the old slow wind illuminated by reflection of light from the buried star.

The Hubble telescope's view of the "Egg
Nebula" gives us a superb glimpse of a rare and
ephemeral state in an ordinary star's lifetime.
A huge Mira variable has lost its outer envelope.
So much of the departing gas has condensed to
dust that it hides the quickly evolving star from
view within the dense inner band. Spreading
outward are two dusty lobes that catch and
reflect the light of the buried star back to Earth.
(R. Sahai and J. Trauger of JPL, the WFPC2
Science Team, STScI, and NASA.)

When the shrouded star's surface hits the magic temperature of 25,000 degrees Kelvin, it produces enough ultraviolet radiation to begin to ionize the thick shell that the violent fast wind had created, and a new planetary nebula enters the Galaxy's catalogues. As the nebula expands and the star heats, the radiation invades deeper into the enveloping cloud, revealing progressively more of its complex structure, and finally penetrates beyond the thick shoveled shell to light up the ancient surrounding wind. After a few thousand years, by which time the star has heated to over 100,000 degrees Kelvin, its protective hydrogen envelope is almost all gone, and nuclear fusion begins to shut down. The star then dims and cools, as a planetary nebula, now grown light years across, fades into the interstellar gloom, the star now a degenerate white dwarf that will someday be free of its surrounding cloud.

White dwarfs in double-star systems give testimony to such enormous mass loss. Sirius A is a class A1 white main-sequence star of 2.4 solar masses. Sirius B, the white dwarf, weighs in at 1.0 solar. But high-mass stars die first. Sirius B must once have been the more massive of the two, a class B dwarf of perhaps 3 solar masses. Some two-thirds of it is gone, no trace left, its planetary nebula evaporated. Sirius A will one day repeat the behavior.

Circling the Galaxy forever, a white dwarf is fated only to cool and dim from hot and blue to chilly and red. If there is a little hydrogen left, the helium will sink out of sight, and we see the star as hydrogen-rich. If it has lost its entire hydrogen envelope, it will be helium-rich. (The matter is unfortunately not so simple. The two kinds of white dwarf show up at different temperatures, as if white dwarfs can change their minds about what kind they wish to be. No one knows how or why.)

White dwarfs are sheathed in a thin, but efficient, blanket of insulating nondegenerate gas. Cooling is slow, so slow that the Galaxy is not old enough for any white dwarf ever to have chilled below the surface temperature at which it would be visible. By finding the faintest white dwarfs and calculating how long they should take to arrive at their stations, astronomers can calculate another age for the Galactic disk. (Halo stars are too faint and sparse for their statistical study to be complete.) It is about 10 billion years, the same number found from the open clusters. We *are* doing something right.

the road to disaster is paved with bright stars

As the mass of a star increases along the main sequence, so does the mass of its fusing core and of its ultimate white dwarf. The Sun will make a white dwarf of around 0.6 solar mass. Sirius B, the progeny of a more massive B star, was left with the mass of the Sun. At 10 or so original solar masses, the white dwarf hits its own magic number, the very special 1.4 solar masses.

In the 1930s, a young Indian astronomer name Subrahmanyan Chandrasekhar fit the equations of Einstein's relativity to those that describe the behavior of a white dwarf. He found that as the mass of a white dwarf approached 1.4 times solar, the degenerate electrons were forced to move so fast in support of the star that they approached the speed of light, at which point the rules change. At this magic number, called the Chandrasekhar Limit, a white dwarf can no longer hold itself up and has no choice but to collapse. No star with an initial mass greater than about 10 solar can become a white dwarf. 10 solar masses correspond to spectral class B0.5. The O stars (and the hot limit in class B, which for convenience will be lumped with class O) must become something else.

O stars begin to evolve much like those of lower mass, cooling and brightening a bit while fusing hydrogen into helium. One difference lies in the winds, which, because of the O stars' great luminosities, are fierce. Almost as soon as they are born, the hotter O stars start whittling themselves away. When the core hydrogen runs out, the evolutionary pace picks up. The stars swell and begin to cool at their surfaces; but unlike the lower-mass stars, they cool at more or less constant luminosity.

Just as lower-mass stars split at around 1.5 solar masses, class F2 or so (when the carbon cycle takes over), the O stars split at around 50 solar, the highest masses (from 50 to 120 solar) falling into the rarefied and hottest class O3. As their helium cores shrink, the "cooler" (!) O stars evolve as increasingly large supergiants through all the spectral classes from O to M, where the central temperatures and densities become high enough to initiate helium fusion to carbon (and oxygen). Here the stars settle down for awhile, helium-fusing cores nested within hydrogen-fusing shells, to make M supergiants like Betelgeuse and Antares. As progeny of the O stars, M supergiants are just as rare, in fact rarer, as they do not survive the state very long. (They seem relatively common only because of their great luminosities, which allow them to be seen over great distances.) At their largest, they approach the size of the orbit of Saturn, the term "supergiant" apt indeed. Even this set has qualities that depend on mass. In the lower range, below around 15 to 20 solar masses, the stars cross the HR diagram's instability strip, and pulsate as longer-period Cepheids. From around 25 to 50 solar, they do not linger as M supergiants while fusing helium, but loop back to higher surface temperatures, back to class B, twice becoming blue supergiants.

All the while the stars are evolving, they are losing more and more of their mass. At the extreme, so much mass can be lost that some stars strip themselves down to where the by-products of

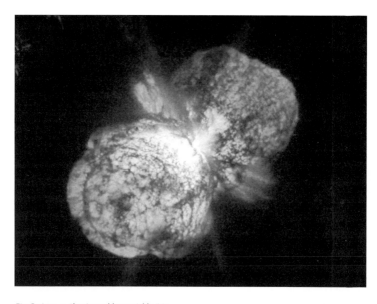

Eta Carinae, a "luminous blue variable," is
deeply hidden within its own dusty ejecta,
seen in exquisite detail by the Hubble Space
Telescope. One of the most massive and
luminous bodies in the Galaxy, Eta Carinae is
losing vast amounts of matter as its core
slowly works its way to iron—and disaster.
(J. Morse, University of Colorado; K. Davidson,
University of Minnesota; STScI; and NASA.)

nuclear fusion appear at the stellar surfaces. The stars then dramatically change their outward chemical compositions, some becoming rich in helium and in nitrogen or carbon, the newly manufactured stuff lost to space through winds. Unlike the solar wind, which is lost through magnetic effects, and Mira-type mass loss, which is driven by pulsations and radiation pressure on dust, massive hot stars drive their winds by means of radiation pressure on the gas. Light exerts a pressure when absorbed by an atom or bounced from a particle like an electron. Radiation pressure even in the core of the Sun is not significant. But at the surface of an ultramassive hot blue supergiant, the outward force exerted on matter by radiation equals or exceeds the gravitational ability of the star to hold it back, and out it goes. Above about 50 solar masses, so much matter is lost that the stars stall in their evolution and cannot swell and cool enough to become red supergiants.

In some high-mass stars, huge outbursts of matter partially condense to dust, hiding the stars within and producing enormous visual variations. The most famed example is the southern hemisphere's fifth-magnitude Eta Carinae, though the star's great nineteenth-century outburst (when it became one of the sky's brightest stars) may have been caused by a now-lesser companion to the visible star. Each of the pair (if indeed it *is* a pair) may have started off as greater than 100 solar masses. Several other of these "luminous blue variables" are known, some in other galaxies. Though they are very rare, because of their great brilliance we have little trouble finding them.

The huge energies pouring from these stars, the vast amounts of mass lost (solar masses nearly at a time), the speeds of evolution, the stars' inabilities to become white dwarfs, all suggest that something quite remarkable is about to happen. It is.

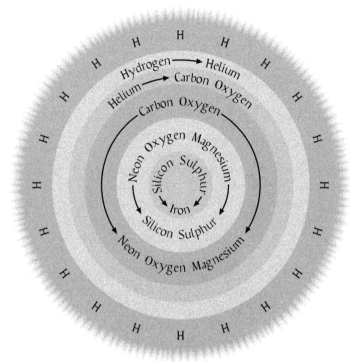

While the cores of lower-mass stars evolve
and burn to only carbon and oxygen, higher-
mass ones keep going through various stages
until a developing iron core is wrapped within
nests of other, previously burning, states (all
surrounded by a vast hydrogen envelope).
Once the iron core is complete, it will collapse
and the star will explode.

catastrophe

The central core of a massive star grows to the Chandrasekhar Limit. But even with immense mass loss, the internal temperatures and pressures keep it from becoming degenerate. Once core helium fusion has subsided, the core contracts and heats, helium fusion spreading into a shell as it does for lower-mass stars. And here is the great divide. Instead of dying as a degenerate ball of carbon and oxygen, the core keeps going. When the temperature climbs high enough, the carbon and oxygen (in this considerably simplified account) start to fuse to a mixture of neon, oxygen, and magnesium, the advanced nuclear fire now surrounded by fusing shells of helium and hydrogen.

Not much total energy is left to be recovered in the march from hydrogen to iron, so "carbon burning" takes a much shorter time than core helium fusion. When it is over, the complex core contracts further, and carbon fusion spreads into yet another burning shell interior to those fusing helium and hydrogen. When hot enough, the neon-oxygen-magnesium mixture fires up and fuses to one of silicon and sulphur, which takes an even shorter period. The process repeats itself when that fuel runs out, the shrinking core now becoming hot enough to fuse silicon and sulphur into iron. Which takes about a week.

For a brief glorious moment, the star hangs in balance, at the edge of a cliff. It has an iron Chandrasekhar-mass core that for the briefest of moments is supported by degenerate electrons, the core surrounded by nested shells fusing mixtures of silicon-sulphur, neon-oxygen-magnesium, carbon-oxygen, helium, and hydrogen all into their heavier ashes. The whole affair is wrapped by what is left of the inert hydrogen-helium envelope, most of which has been lost to winds. Iron cannot fuse to anything to produce energy. Electrons smash into the iron atoms, breaking them down, and are swallowed by protons to make neutrons and neutrinos. Though it has taken the star millions of years to reach this state, with nothing left to stop the collapse and the absorption of electrons, evolution now accelerates to blinding speed.

In under a tenth of a second, the massive core drops from a sphere the size of Earth to one the size of a small town. The density climbs so high that all the nuclei—the protons, neutrons, and electrons—are smashed together to make nothing but neutrons. At a density above that of nuclear matter, over 10^{14} grams—100 million tons—per cubic centimeter, not only are the neutrons highly degenerate, but the strong force that binds atoms together becomes repulsive. As fast as it started, the catastrophic collapse halts. Like Heaven's basketball slammed into Hell, the core violently rebounds, and sets up an outbound shock wave (an ultrapowerful "sonic boom") that is destined to blow the rest of the star apart. Here are the Type II, *core collapse,* supernovae. For an instant, at the moment of collapse, the rate of energy release (including the neutrinos) is comparable to that of all the stars in the visible Universe.

We actually watched it happen. In 1987, the closest supernova since Kepler's star of 1604 erupted in the Large Magellanic Cloud. Though 150,000 light years away, it grew to third apparent magnitude, probably the most distant single star ever viewed with the naked eye. Observers caught the star, formerly a class B1 supergiant, as it visibly brightened, so we knew when the shock wave hit the star's surface. (Even at a shock wave's great speed, it takes time for it to move through an exploding star.) Three hours before the star brightened to the eye, about a dozen neutrinos hatched by the electron-proton mergers smashed into the neutrino detectors in the United States and Japan. Given the distance and the neutrino pick-up rate, the right number hit at just the expected time.

Now go back to the budding supernova's shock wave. As it forms, matter continuing to fall inward pushes against the shock and keeps it from expanding. It stalls. And here, in the words of one prominent theoretician, "is where the shouting starts." Clearly, since supernovae exist, the shock eventually gets out. No one, however, quite knows how. One camp suggests that the shock does the job all by itself. Another invokes the abundant neutrinos. Neutrinos will normally pass through anything; half a light year of lead would have difficulty stopping one. But the density inside the star is so great that the neutrinos are trapped for several seconds. Their absorption heats the gas behind the shock wave and releases it to roar outward. A third view is that the collapsed core turns over in violent convection, thereby bringing up energy from the core's interior, which shoves the shock outward. Such is science, such the search. Eventually, through improved theory and observation, one of these pictures, or none of them, will attain ascendancy.

Temperatures are so high in the supernova's exploded debris that nuclear reactions create all the chemical elements, including a good fraction of a solar mass of iron; these newly-minted elements, as well as the ashes of millions of years of fusion, are all blasted into space. All the iron of a fence, of a car, all the aluminum, all the silicon of a beach, are gifts of aeons of supernova explosions that infused themselves into the interstellar gases before the Sun was born. Only in this most intense of all fires can many of the elements be forged.

Slow neutron capture in giants creates most of our zirconium, our barium, and large fractions of other elements. In this exploding maelstrom, however, neutrons are captured not slowly but rapidly. In the slow neutron capture process, an atom is built through its higher isotopes until a radioactive one jumps to the next element above. If the neutrons hammer away fast enough, however, another will be caught before the jump can take place. *Rapid neutron capture* will therefore build quite heavy isotopes in spite of short radioactive life-times, until one is reached that is so very unstable that nothing can hold it back. In a rush it jumps not to the next element number, but over several element numbers at once to make all our uranium, almost all our silver and gold, and portions of many other heavy elements.

What is the Earth made of? Iron, nickel, silicon, and oxygen, and not too much else. Nearly all we have of these elements came from supernovae, but much came from lower-mass and less spectacular evolution, too. When theoretical astronomers calculate the expected abundances of the Universe as generated by stellar evolution, they find something quite akin to what is found in the Sun. When they remove the hydrogen and helium, they find what is in the Earth. We are the distillate of supernovae. We live in an extraordinary place.

the end: part II

The blasted debris of a supernova is blown outward at 10,000 kilo-meters per second or more, creating a *supernova remnant* like the Crab Nebula, as well as 200 other remnants seen within the confines of the Galaxy. The gas, rich in heavy elements, deposits the load into the dusty interstellar gases to await the formation of new stars. Long after the original envelope of the star has disappeared, we can still see the blast waves penetrating, sweeping up, heating, and compressing the interstellar medium, where it helps create new stars. Riding the shock wave like minuscule surfers are vast numbers of protons and other atomic nuclei, accelerated to nearly the speed of light. The charged particles —electrons too—are trapped by the Galaxy's magnetic field. Striking Earth, these "cosmic rays" smash atoms in the upper atmosphere, the debris raining to the ground all around us, all

In eastern Cygnus lies a shell of gas some 2 degrees across (4 times the angular diameter of the full moon) heated by the still-expanding shock wave from a star that exploded some 50,000 years ago. The expanding arcs (the eastern one seen here) consist of complex, heated filaments that show the violence of the great event. Such shock waves heat the interstellar gases and also compress them to initiate star formation. (© Malin/IAC/RGO. Photograph by David Malin.)

unseen. Exploding stars reach out not just with their light, but with their fingers.

Sit in a revolving chair and stretch your arms. Have someone spin you. When you pull in your arms, you spin faster. Watch a skater do the same to propel a spin. In an orbiting body, the product of mass, radius, and velocity is the body's *angular momentum*. When you spin, your angular momentum is the total of the angular momenta of all your parts. In the absence of an outside force, total angular momentum cannot change. When you bring in your arms, your average radius goes down, so your speed goes up. The core of an evolving supergiant is rotating along with the rest of the star. Collapse it to a ball of neutrons and we see a neutron star that is spinning very fast, perhaps 30 or more times per second. Its magnetic field collapses, too. Once the exploded outer envelope of the supernova has dissipated, the neutron star—now a pulsar—is revealed, as it was for the Chinese Guest Star of 1054, which created the Crab Nebula and the Crab pulsar.

The power of pulsar formation is shown by the pulsars' motions. As stars of the Galactic disk, O stars and supergiants do not move fast relative to the Sun. But pulsars are often seen to be streaking away, fast enough to leave the disk. The detonations of supernovae are off-center, and are so strong they can give the resulting pulsar—an entire star!—such a kick that it can leave the Galaxy's disk, where the massive stars were formed. We have no idea why the explosions are not centered.

Theories now become even murkier. The celestial X-ray source called Cygnus X-1 appears to contain a black hole. Presumably, it is the remains of a high-mass star. As white dwarfs have a limit to their support by degeneracy, so do neutron stars. Harder to calculate, the limit lies between 2 and 3 solar masses, not much greater than neutron stars found in double systems in which their masses can be measured. The invisible companion in Cygnus X-1 and in the other candidates well exceed the limit, powerful evidence that black holes are truly there.

We surmise that only the heaviest of O stars can be parents to black holes, that only these can make cores big enough. But how the process works is still a mystery. Indeed, how supernovae can be created along with black holes is a mystery: with no degenerate neutron star and rebounding core, how is the driving shock produced? Quite possibly there is no shock, and the star simply falls into the black hole, a "fizzler" instead of a supernova. Another possibility arises from the observation of the distant gamma ray bursts, which dwarf the power of "ordinary" supernovae. One was identified with a supernova a hundred or more times brighter than the "ordinary" kind, a *hypernova*. When a massive stellar core collapses to make a black hole, it may still be surrounded by large amounts of spinning matter. The rapid infall of this debris into the black hole immediately after formation may multiply the power of the collapse, turning "super" into far grander "hyper."

and what about double stars?

Core collapse can take place only in massive stars that inhabit the Galaxy's disk. It cannot produce the Type Ia supernovae found in galaxy halos and in galaxies with no young-star populations. Where do such supernovae come from? White dwarfs in double systems make novae. Why not supernovae, too? Again, there are several viable theories (which seems odd, since astronomers rely on these supernovae to gauge distances to far-away galaxies, and perhaps should first be sure they actually understand them).

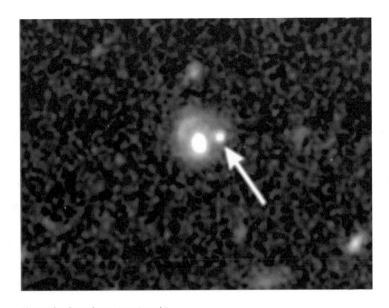

The transfer of mass from a star onto a white dwarf can cause the tiny star to explode so violently it can be seen over distances of more than a billion light years. The uniformity of such outbursts allows the measure of the expanding Universe. (P. Garnavich, Harvard-Smithsonian Center for Astrophysics; the High-z Supernova Search Team; STScI; and NASA.)

Double-star evolution goes off quite in its own direction. Start two stars together, not too far apart, but not so close that they initially have much effect on each other. The more massive evolves first. As it grows, it encroaches upon its zero-gravity tidal surface, and mass flows from one star to the other: Algol! If the two are close enough, the flow from the more massive might even encompass the pair in a common envelope, one star actually orbiting inside the other. Friction makes them spiral together. When the evolution of the first star is over, a white dwarf may find itself tucked up close to an ordinary dwarf on the main sequence. Again, if close enough, the white dwarf tidally distorts the ordinary dwarf, mass flows onto the white dwarf through an accretion disk, and, with enough time, a nova erupts.

Now what if the white dwarf is the offspring of a fairly massive star and is right at the edge, near the Chandrasekhar Limit? The new mass flowing from the ordinary dwarf may push it into the white dwarf netherworld, in which it can only catastrophically collapse, the energy release blowing the star apart, maybe annihilating it altogether, and making a supernova even brighter than the core-collapse variety. Type Ia supernovae can then appear anywhere.

Another possibility awaits further stellar evolution. A white dwarf quietly orbiting in a double system eventually sees its originally lower-mass partner begin to swell. As a cool giant, the partner grows large enough again to pass mass to the white dwarf. The heated flow produces features in the spectra that no giant star should have. Such stars, called *symbiotics*, abound, some involving Mira variables. Again, a common envelope could draw the stars together, the eventual result a double white dwarf in close gravitational embrace.

Now invoke esoterica. Relativity predicts that orbiting bodies should make gravity waves that carry off energy, the two thus slowly spiralling together. If they draw close enough, they merge. If the product is over the Chandrasekhar Limit, they collapse; they explode and disappear, the light so bright they can be seen more than a billion light years away.

Doubles may also be responsible for gamma ray bursts. We know that neutron stars and black holes reside in binary star systems. Given the power of a white dwarf supernova, imagine the violence generated by the mergers of neutron stars with each other—or with black holes. Theories to explain these distant, violent gamma ray events rain down on us almost as fast as the bursts themselves.

Such mystery is hardly disconcerting. After all, science is a search for that which we do not understand. As more evidence pours in, someday we will. When the twentieth century dawned, no one knew that there were such things as supernovae. We had no idea of what kind of stars were in the sky, how they came to be, or how they died. Looking back a hundred years hence, we will almost certainly have learned the answers to the questions of today. Perhaps some young person drawn outdoors by the captivating power of the nightly parade will find them.

dawn

Observation tells us what kinds of stars there are, and theory tells us how one kind becomes another and what finally happens to them. But where did they all come from in the first place? We know stars are always dying, yet there are still huge numbers of them out there, and because some stars are clearly very young, the birth process must be continuous. How and where does star birth happen?

To find out, we apply the same logic to star birth as we do to star lives and star death. We look for clues that might imply that stars are new, then use theory to connect them. But here a severe problem rises, like a Scottish mist shrouding the landscape. Our eyes alone reveal little: we do not see stars suddenly pop into being, nor do we see anything very dramatic that might suggest stars in the immediate process of formation. Their births must be hidden.

Where is a good hiding place? Look again into the Milky Way, look for the dark spots, the black clouds in which there seem to be no stars. They cannot be real gaps, as stellar motions would fill them in. Moreover, they are frequently adjacent to bright clouds of interstellar gas in which young high-mass stars are embedded. As early as 1922, astronomers Henry Norris Russell (of HR diagram fame) and Bart Bok (a Dutch-American astronomer known for his work on the structure of the Milky Way) suggested that the dark clouds must be stellar breeding grounds, and that something, some obscuring material, was keeping us from seeing what was going on within. They were right. Proof, however, had to await advances in technology that gave us the ability to penetrate the clouds: the twin cannons of radio and infrared astronomy.

The dark clouds, we now know, are made of cold gas mixed with fine dust. The tiny grains within are efficient absorbers of visual radiation. But longer wavelengths penetrate beautifully, just as radio broadcasts remain undiminished on a foggy day. With these tools of detection, there they are, all manner of strange objects that can only be stars in various stages of inception. Theoretical analysis of these obscured stars then allows us to identify visible but odd stars as new. Star birth, no longer hidden, allows us to make the full tour from birth to life to death and back to birth again as the stars recycle themselves, one generation helping to create the next.

in the clouds

Darkness, everywhere darkness, achingly cold darkness. Against the background of the Milky Way are black blots—"globules"—gas clouds so dense with dust that no starlight whatsoever seeps through from the other side. Were the Sun and our Earth within such a cloud, we would see no Milky Way, no other stars at all, possibly just a few nearby ones that were in the same fix. Passage of the Sun through a dense globule might even affect the heating of the Earth and cause climate changes. Not mere speculation, this, as the Sun has indeed been inside such a cloud at least once: at its birth.

What would we have witnessed? How were the Sun and all the other stars born? What events conspired to bring light from such utter gloom, to raise dawn from the blackness of the dusty Galactic night? The clouds range in size from a few light years across to hundreds, in mass from many Suns to hundreds of thousands. To understand the clouds, first look outside them. Interstellar space is filled with a thin, lumpy, hydrogen-helium gas (in the same ratio as found in the Sun, no surprise), an *interstellar medium*. Much of the hydrogen is ionized by energetic ultraviolet light from distant stars, the temperature some 10,000 degrees Kelvin or so. (The gas does not radiate much in spite of its high temperature because it is so vacuous, rather like the Sun's corona.) Within the gas are the ubiquitous flecks of dust that originate mostly from Mira variables. The grains, so small you could not see them, together constitute about 1 percent of the medium's mass.

The grains' effects, however, are all out of proportion to their size. Interstellar grains scatter and absorb starlight. As a result of the absorption, stars (at least distant stars in the Milky Way) are fainter than they should be, some by many magnitudes. Interstellar dust can drive the astronomer mad, as it makes distant stars (those beyond parallax) appear too far away. Until the presence of the dust was accounted for, we grossly overestimated the size of our own Galaxy. There is, however, a panacea. Blue starlight is extinguished to a greater degree than red starlight, so stars are also reddened by the grains, such that their colors are not consistent with their spectra. Since reddening (which is measurable) and total absorption are

Intensely dark clouds fleck the Milky Way.
They are filled with lumps of cold dust and
molecular gas, just the conditions needed to
form stars. A look inside with radio or infrared
telescopes often shows stars in the making.
(© Anglo-Australian Observatory. Photograph
by David Malin.)

coupled (one increasing in lockstep with the other), we can derive the latter from the former, at least most of the time (the exceptions from odd dust driving the astronomer even madder).

Now reenter the dark clouds. The gas density of the heated medium is a mere atom per cubic centimeter or less. In a cloud, it goes up by factors of thousands or more, as does the density of the dust grains. Heating starlight is eliminated, so the temperature drops to only a few tens of degrees Kelvin, in some places to near absolute zero. With no starlight, the gas becomes neutral, and with the grains, sets up a symbiosis of sorts that results in a complex, cold chemistry. Reactions are slow, to be sure, but we have all the "lab time" we need. A cold, slow-moving hydrogen atom sticks to the surface of a grain, then another and another. Finally, a hydrogen atom hits one of the passengers, joining it to make a hydrogen molecule that is kicked off by the energy released in the reaction. The dark clouds are therefore dominated by molecular hydrogen.

Starlight, which would destroy the molecules, cannot enter the clouds. But cosmic rays, ejected by supernovae, can. A proton moving at nearly the speed of light streaks through the dusty gas, leaving a thin trail of ionized hydrogen molecules behind it, and we are on our way. Ions will readily join together. Combining first with itself and then with heavier atoms, the molecular hydrogen forms a base (along with grain-surface chemistry) to build the more than 100 molecules now known, and surely many more. These are no longer just "clouds": they are *molecular clouds.*

Since we cannot see into the clouds, visual observations tell us little. Instead, the cold molecules both absorb and emit low-energy radio photons that are observed and analyzed with radio telescopes, each kind of emitting molecule acting like a radio station broadcasting on different multiple frequencies. The radio astronomer simply "tunes the dial," compares the signals with those observed for different substances in the laboratory (or with theoretical prediction), and announces the discovery of yet another molecule. Present in the clouds are carbon monoxide, water (lots of it), alcohols, various acids, hydrocarbons (methane and acetylene), what we would call paraffin derivatives, unstable things not found on Earth, and more. Much of this chemical mix, along with individual heavy atoms, condenses onto the grains, making them far more complex than they were when they left their parent stars. On the top rung of the ladder stand an amino acid (glycine) and huge linked benzene rings, these approaching the sizes of the grains themselves. We have no idea how far the chemistry extends. But amino acids are the foundations of proteins. This cold chemistry may in fact provide the foundation of life.

it takes a galaxy to raise a star

To understand star birth, we must first construct the birthing clouds. Like the early American West, the wide-open spaces are violent places. Interstellar gas is in a state of constant turbulence. Look at the spiral arms of other galaxies. They are not permanent, but are waves of density that propagate through the fields of stars and through the gassy medium, compressing it. Rotate the galaxy, and the waves spread out into winding arms. The Sun, in circular orbit, keeps passing in and out of those that belong to our own system.

The compression is erratic, the gas clouds shredding into smaller clouds and shards that collide, merge, and shred again. The interstellar medium cannot possibly avoid becoming tattered. Supernovae from previous generations of stars add to the mess, their driving shock waves compressing the gas even more, the shocks so strong that they make tiny diamonds (observed!) from amorphous carbon grains. The result is clouds of all sizes and shapes, from tiny knots to giant molecular clouds that contain up to a million times the mass of the Sun, making them the Galaxy's largest structures.

Under such a state of siege, not only will the interstellar medium be in a state of extreme lumpiness, but so will the individual molecular clouds. Spotted here and there within them will be ultradense blobs perhaps a light year across. The blobs are so cold that gas pressure within them is low. If they are massive enough, their own gravity can take over and make them contract. Ammonia within high-density lumps like these becomes an efficient radiator. When we tune our radio telescopes to ammonia's frequencies, there they are, *dense cores* buried in the molecular clouds, our first look at stars in the making.

Everything, even molecular clouds, rotates. Bumping and smashing into one another (if only in the gravitational sense), how could they not be set spinning? The dense cores are no exception. Allow a small one to contract, to collapse under gravity. As its radius becomes smaller, it rotates faster, the phenomenon no different from that which creates a pulsar from a collapsing iron core. As the dense core's spin speeds up, it flattens, and finally rotates so fast that it tears itself apart. Stars can therefore not be born. End of chapter.

But no. Obviously they *are* born, otherwise we would not be here. The problem is, how do they lose their spin? Slowing down a core a light year or more across would seem like stopping an enraged elephant by hand. However, with a remarkable synergy that involves not only star life and star death, but the entire Galaxy, it can be done. The Galaxy rotates and is filled with ionized gas. As a result, it creates a magnetic field about a millionth the strength of Earth's. (As weak as that seems, totaled over the Galaxy, it represents a huge amount of energy.) This pervasive magnetic field threads through the molecular clouds as it does everything else. The ions made by cosmic rays (the same that aid in cloud chemistry, produced by the same supernovae that compress the clouds in the first place) are few, but enough to grab onto the magnetic field. The ions thus feel a force that attempts to bring them to a halt. Collisions between the ions and the neutral atoms and molecules transfer the force to the dense core as a whole, which slows both its contraction and rotation. Gradually, the infalling neutrals escape the ions and the field, but by then the collapsing blob is spinning much more slowly, slow enough perhaps to form a star.

Or maybe not yet; it may still tear itself apart, not completely, but into two lobes. Spin energy now goes into orbital energy, the individuals rotating considerably more slowly. Each of these may now condense to make a double star. If one is still rotating too fast, it—or both—may again bifurcate to make a triple or quadruple star, explaining why we see so many doubles and double-doubles in the sky. Indeed, the very existence of doubles and double-doubles tells us something of what must happen.

None of these processes can actually stop the rotation; there must always be some. The idea is to remove enough rotational energy so that the final product can collapse into a "sun" without disruption. As it condenses, the internal pressures and temperatures build toward the point at which thermonuclear reactions can begin. At the same time, matter is still falling inward from the parent cloud. As the nascent star rotates faster, the outer parts of the collapsing mass should still spin out into a dusty disk. The disks were the keys to discovery.

In the 1940s, optical astronomers began to find what eventually led to lists of hundreds of distinctive erratic variable stars in the neighborhoods of dark interstellar clouds. The stars' modest temperatures placed them in spectral classes F to M, but they were notably more luminous than ordinary dwarfs, falling on the HR diagram between the main sequence and the giants. Their spectra, however,

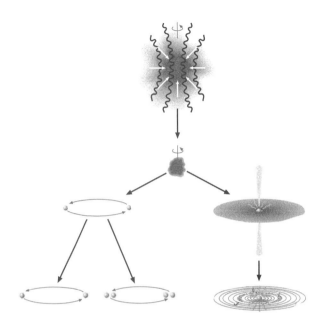

A dense cloud of dusty gas, supported for a
time by the Galaxy's magnetic field (wavy
lines), which also slows the cloud's rotation,
begins to collapse to form a star. If rotating
slowly enough (right), a single star forms, the
residual rotation making a disk and flowing
jets that carry away still more rotation. Planets
can then form within the disk. (*Orbits are
not to scale.*) If the star-to-be is still rotating
too fast after its magnetic braking (left), it
might split to form a double star, which can
split again to make a double-double. Other
combinations are quite possible, as are disks
around the individual binary components.

looked like neither. Emissions of various elements revealed the presence of surrounding low-density gas. Excess ultraviolet light suggested activity far greater than seen in the Sun (that associated with the solar magnetic, or sunspot, cycle).

These *T Tauri stars* (named after the prototype) were also found to radiate much more infrared light than expected for normal stars, radiation that could only be coming from heated, surrounding dust. There is in fact so *much* dust that all the visual light should be absorbed, the stars completely hidden from view. Yet there they were, quite visible to the eye. The only way we should see one is if the dust were distributed in a disk, one flat and thin enough so that the star peeks out over the edge. Where there is dust there is gas. The intense ultraviolet radiation, as well as the irregular variability, strongly suggests that matter is raining down onto the stars from the disks, that the disks are in fact *accretion* disks.

Emerging from many of these stars are opposing flows and jets that can extend for light years before they sweep up enough ambient interstellar matter to bring them to a halt. The result is pairs of bright clouds (*Herbig-Haro objects*, named after George Herbig of Lick Observatory and Guillermo Haro of the University of Mexico) with the young stars in the middle. The jets from such a star almost certainly have to be focused through the poles of a disk. Radio astronomers also found similar distinctive flows buried deep within the obscuring dusty molecular clouds. At their centers were dense bright ammonia-radiating cores. In some flows, the ammonia radiation came from disks, firmly connecting these dense cores with T Tauri stars, which must then be the cores' successors. Detailed pictures taken with the Hubble Space Telescope (as well as other instruments) reveal the disks themselves, as well as the details of the outbound flows. The flows, which carry away additional spin energy, act as giant fingers pointing toward new stars. As matter falls onto a new star from its accretion disk, forces not yet well understood—perhaps magnetism generated by the spin—begin to eject matter in the perpendicular direction. As the rate of infall slows, the flows die away, leaving a T Tauri star embedded in a quiet, circulating, orbiting disk.

As mass accretion proceeds, the stars first become very hot and bright through simple compression. When hot enough inside, they begin to fuse the natural deuterium given them by the Big Bang, and hang above (the stars brighter than) the main sequence. Here is the true "moment of birth." From there, as mass accretion dies away, the stars contract at constant surface temperature, and therefore dim to the eye at the same time that the interior temperatures are climbing. As deuterium fusion diminishes, the stars' structures change. They stop dimming and begin to heat at their surfaces. When the core temperatures hit just short of 10 million degrees Kelvin, full fusion, the proton-proton chain, kicks in (and if there is enough mass, eventually the carbon cycle). The stars quickly settle into their initial positions on the main sequence, from which they will, ever so slowly, age. We have found stellar birth, have found the new stars, and by theory know how they are born. And it took the whole Galaxy to make them.

tops and bottoms

The lumps within the molecular clouds come in all sizes, and so do the stars. Some clouds spawn not just stars, or even multiple stars, but whole clusters. If the molecular clouds are large enough, they may be birthing several clusters at one time. Moreover, some of the stars will be large, too, a small few collecting the vast masses of O stars. The highest mass O stars contain, as far as we know, somewhat over 100 solar masses; the limit is imposed by huge radiation pressure that tears high-mass stars apart. Simple statistics also militate against

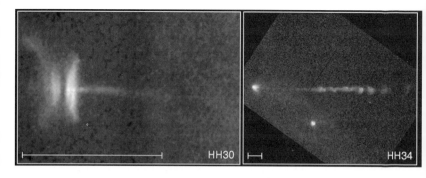

The Hubble Telescope clearly images the flowing
jets and surrounding disks of developing stars. At
left, the disk hides a star that illuminates its disk's
edges. Are planets forming within? (C. Burrows,
StScI; J. Hester, Arizona State University; J. Morse,
StScI; and NASA.)

high-mass stars. There is not enough mass in molecular clouds to create many of them, and beyond a high-mass cutoff, the odds of making them simply vanish.

O stars, all above 25,000 degrees Kelvin, can ionize the surrounding gases of the molecular clouds that gave them birth, destroying most of the molecules within and producing bright *diffuse nebulae* that behave like planetary nebulae but are vastly larger and more massive. Diffuse nebulae abound in the Milky Way at the fringes of dark molecular clouds, where they are not so obscured that they cannot be seen. They range from tiny dense knots to vast structures lit by entire clusters of O stars. Among the most beautiful and photographed of all celestial sights, they are, above all, markers of star birth. Some are so bright they are visible to the naked eye. All, big and small, beckon: "Here are new stars; come look."

Chief among the diffuse nebulae is the great Orion Nebula, which surrounds the middle star of Orion's sword, a stunning sight even through binoculars. With the telescope, we see vast streams of shredded gas overlain with foreground dust clouds. Buried inside are four O stars—the Trapezium—the brightest of which does most of the work in ionizing and heating a hundred solar masses in a cloud several light years across. This quartet, which is gravitationally unstable and will break up, is at the top of a huge cluster of fainter stars that pervades the nebula. In relief against the nebula, the Hubble Space Telescope has spotted numerous faint new stars surrounded by—what else—disks of circumstellar matter.

Opposite, in Sagittarius near the winter solstice, lies another, the Lagoon Nebula, one among many within this and neighboring constellations. Within the Large Magellanic Cloud shines the great Tarantula Nebula, so big that if placed at the position of the Orion Nebula, it would fill the entire constellation. Hundreds are cata-

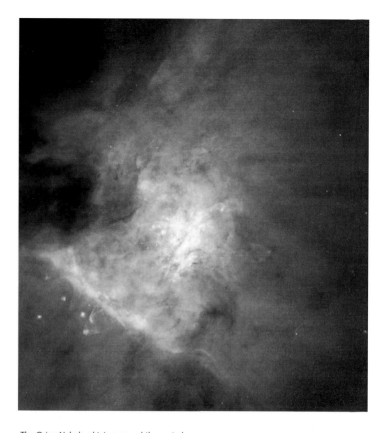

The Orion Nebula, shining around the central star of Orion's mighty sword, is lit by massive new stars and contains thousands of lesser lights. Behind it is a massive molecule-filled dusty cloud that is birthing yet more stars. (C. R. O'Dell and S. K. Wong, Rice University; STScI; and NASA.)

logued, even those that are deeply buried within their parent clouds and observed only by their escaping radio radiation.

A broader look shows great groups of high-mass O and B stars moving more or less together through space. Too widely separated to be bound by their own gravity as clusters, these *associations* are spreading apart. The high-mass stars, however, do not get very far away before they burn out, the highest masses producing supernovae that will spawn more star birth and more associations in a self-perpetuating cycle. The nearer associations make major parts of whole constellations: Orion's sparkle comes from associated O and B stars; Scorpius is the product of another such structure, Perseus of yet another. As one association burns out, another takes its place, the lower-mass stars born along with them living long enough to run free, those in actual clusters staying there until the tides of the Galaxy finally rip them away.

Nature must love lower-mass stars. For every O star, she makes a million cool M dwarfs. Life at the bottom is hardly illustrious, but the stars are compensated by long life and a great many neighbors. The lower limit of 0.08 solar mass is imposed by the ability of the star to run the full proton-proton chain. But that is not really the end. Beneath the bottom is a grey limbo. There is no reason why the cloud condensation mechanism should respect any fusion limit, and there should be failed stars below it, bodies that have condensed with masses below 0.08 solar, the brown dwarfs.

Brown dwarfs are cool, faint, and hard to find. They live off gravitational contraction and from fusion of natural deuterium. With temperatures that descend from around 2000 degrees Kelvin, they radiate most of their light in the unseen infrared, where they can be seen in abundance with infrared detectors as stars of new class L. At the low extreme are a few bodies (those of spectral class T) that have methane absorptions in their atmospheres, something no respectable star could have, the temperatures hovering at only 1000 degrees Kelvin. The real bottom to the masses of the droplets rained by nature's interstellar clouds, and the numbers of them, are unknown.

meanwhile, back in the disks

The first evidence that new stars should have disks was recognized over 200 years ago, when Immanuel Kant (and others) looked at the Solar System and wondered how the planets—and the Earth—might have been born. The planetary system is remarkably regular. All the planets orbit the Sun in the same direction, that of solar rotation (counterclockwise as viewed looking down from the north), and very nearly in the same plane, close to that of the solar equator. Orbital sizes increase in geometric proportion. Most planets also spin in the same direction, with their axes generally perpendicular to their orbits. With some exceptions, the satellites of the planets go around their parent bodies in the same direction too.

Physical relations are equally striking. The inner four planets (Mercury, Venus, Earth, Mars), all within 1.5 AU of the Sun, are small. (Earth, the largest, is only a hundredth the size of the Sun.) They are made of rock and iron and contain very little light stuff and almost no water. (Though the oceans seem vast, they are small compared with the whole Earth.) These "terrestrial" (earthlike) planets increase in size from the Sun to the Earth, then decrease to Mars. Outside the Martian orbit, approaching great Jupiter, are myriad smaller dense bodies, the famed asteroids, of which Ceres (1000 kilometers across) is king.

Now take a great leap. Jupiter and Saturn, orbiting the Sun at 5 and 10 AU, are huge, ten times the radius of Earth. Jupiter is over 300 times Earth's mass (but still only a thousandth the solar mass). Both are made of hydrogen and helium (most of it in the liquid state) overlying deep rocky cores. The next two (Uranus and Neptune, at 19 and 30 AU) are intermediate in size between the giants and the Earth, and are much richer than Jupiter in heavier substances, probably mostly water, which is everywhere, the satellites of the outer planets up to 70 percent ice. At the end, beyond Neptune, are huge numbers of small bodies, of which Pluto (40 AU away and less than half the size of Mercury) is largest. Finally, entering the inner Solar System in long looping orbits are the comets, dirty melting iceballs that glow and send out long tails under the action of heating sunlight and the solar wind. There must be huge numbers of them in the frozen wastes beyond the planets.

There is no way such a system could have been created randomly, piecemeal. The Solar System *must* have been made as a whole, all at one time as part of the formation of the Sun. The temporal relation is now proven, what with the Moon, meteorites (small Earth-hitting asteroids), and the Sun all measured to be the about the same age, 4.5 billion years. Kant suggested that the planets were formed from a spinning disk of gas, the "solar nebula." Understanding this system, in which planets are natural offshoots, by-products, of the formations of stars, is crucial to making sense of the Sun and of star birth. The planets are prime evidence of what happened so long ago.

Only in our own time were we able to find some kind of proof of Kant's claim, proof among the other stars, that young stars are commonly possessed of disks. Kant and others originally thought that turbulence within the gas would ultimately lead to its condensation as planets. The origin of the Earth and its siblings, however, lies not in the gas but in the dust, dust that was responsible for the formation of the Sun in the first place, dust that had its origin in giant stars. Some of the dust and molecular gas in the solar nebula came directly from the parent molecular cloud, but much of it could have condensed and formed during the development of the dense core and disk. As it did in interstellar space, the dust continues to evolve, collecting atoms and molecules from the gas. But now it is in an environment that is heavily influenced by the forming Sun.

Near the Sun the gas is hot, close to 1500 degrees Kelvin, temperatures dropping into the tens of Kelvins in the disk's outer regions, tens of AU away. In the warmer regions, only those elements with high evaporation temperatures can condense. Farther away, elements with lower critical temperatures add themselves to the grains, so the dust's chemical composition changes with distance. At around 4 AU, the temperature hits a crucial milestone, the condensation point of water; huge amounts of the stuff are absorbed by the grains. Only in the most distant regions can really volatile elements be added. The significant result is that near the Sun, the grains are bone dry; in the outer disk, beyond 4 AU, they are watery.

Return to those ancient times. As the planets do today, the tiny dust grains circle the Sun. Subject to friction with the gas, they are slowly dragged inward, their orbits crossing each other. Some gently collide, and electrical forces make them stick together, resulting in larger grains. These also collide, and pick up smaller grains as well, the particles ever so slowly growing yet larger. Before long they are not only visible to the eye (were there anyone to see them), but are growing to millimeter, centimeter, even meter size, some passing the kilometer mark. At this point they are hardly dust grains, but *planetesimals*, the word coined from "infinitesimal planets," "little ones." As they grow, their gravity becomes important, and the big ones quickly gobble the small ones, the planetesimals becoming *protoplanets*, countless thousands of them up to hundreds of kilometers across. The inner ones, those near the Sun, are dry, made of silicates and metals. Farther out, they incorporate huge amounts of ice.

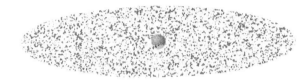

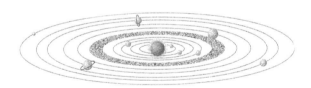

A dusty disk (top) surrounds a new star,
perhaps the early Sun. Its dust grains (middle)
accumulate into trillions of small planetesimals
that keep colliding and fusing until they make
a set of planets (bottom). *(The planetary
orbits are schematic, and not to scale.)*

As the planetesimals and protoplanets increase in size, so does the violence of collision. In the meter or so range they hit each other so hard that it is difficult to see how they can merge, but merge they must. At the extreme, when the larger ones collide, the energy of motion of the colliding body is deposited as heat. The larger body becomes so warm that it turns plastic, even begins to melt. Atoms and molecules then begin to "differentiate," as heavy stuff sinks to the center and light materials float to the outside. Larger bodies therefore take on layered structures, with iron cores (relatively heavy) and silicate exteriors (relatively light), just as geologists find in the Earth and other inner planets.

In the inner Solar System, the limited supply of condensable raw material keeps the planets relatively small. Their gravity is not enough, and they are too hot from sunlight to capture and hold much of the disk's light gas. But in the outer reaches, where the temperature is lower, protoplanets can not only incorporate volatile substances such as water, but can grow large enough to capture and hold vast amounts of hydrogen and helium from the solar nebula. Jupiter and Saturn thus grow immensely fat. Still farther away, where spaces are larger and the raw material more thinly spread, Uranus and Neptune grow more slowly. At the same time, the Sun is turning on and beginning to sweep gas and dust from its disk. Uranus and Neptune cannot therefore accumulate enough light gas as their huge rivals, so come out smaller and denser. Beyond Neptune the disk is too thin; the developing bodies remain as billions of icy planetesimals and protoplanets in the extended *Kuiper Belt* (after the Dutch-American astronomer Gerard Kuiper). The four larger planets develop satellite systems, perhaps in the same way, by spinout from a disk, or by some other process. Near the end, a rival, but smaller, planet hits us, part of it adding to Earth, some vaporizing and reforming as our Moon (the only theory able to account for lunar characteristics). And then the whole affair is over. The planets have been assembled in a mere 100 million years, the figure supported by the dating of the Moon, which is only 75 million years older than the most primitive bodies that fall to Earth as meteorites.

A small planet like ours surely would have formed beyond Mars, and one or more certainly tried, but giant Jupiter got in the way. Its gravity kept the planetesimals there so stirred up that not only could they not accumulate, but they began colliding with such violence that they broke each other apart, the larger differentiated ones producing individual chunks of rock and iron. Now only planetary flotsam orbits in the lonely space between Mars and Jupiter. (Mars probably came out small for much the same reason.)

The planet-making process was hardly 100 percent efficient. Huge numbers of unincorporated planetesimals roamed the system. The inner debris was—and still is—rocky-metallic, the outer icy. Scattered by the planets and crashing into the cooling rocky planets and outer satellites, the pieces of debris dug amazing numbers of impact craters. The record of those early days is still seen on the face of the Moon and on all the other bodies that did not wear them away by "geologic" or erosional processes (as happened on Earth).

The four large planets also scattered vast numbers of the outer ice balls. Some exited forever, but trillions more remained in a loosely bound cloud (the *Oort Comet Cloud*, after the Dutch astronomer Jan Oort, who hypothesized it) that extends perhaps halfway to the nearest star. These, as well as the icy planetesimals of the Kuiper Belt, leak back in as a result of gravitational forces (including those imposed by passing stars and molecular clouds, and by Galactic tides), where they are seen today as comets. Movement of the Sun through a Galactic spiral arm should bring additional cometary rain and collisions. We—the Earth and humanity as well—are very much a personal part of the Galaxy.

Jupiter's gravity also still sends large numbers of asteroids into the inner Solar System. A thousand or so tiny ones hit the Earth each day, falling to the ground as meteorites. Made of either stone or iron, they tell of the heating and separation of elements in the asteroids, and that they are the smashed debris of once-larger bodies. On rare occasion a big one will make it in and dig a hole in the ground, these and cometary collisions reminding us of the early days and convincing us that we are one with the system, that we are one with the stars.

The hot early Earth should have had no, or very little, water. What our planet does have was most likely brought here by infalling icy comets (something to think about when having your next cup of coffee). Meteorites, and almost certainly comets, have a rich molecular mix. Meteorites contain interstellar grains from before the Sun was born, diamond dust from interstellar space, and amino acids. Perhaps the seeds of life were also brought here as molecules made in interstellar space or made in the solar nebula by similar processes, tying us ever closer to the shining lights of the nighttime sky.

anyone else out there?

If the Sun's disk formed planets, why not the disks that surround other stars? That they likely do is demonstrated by the observation that the disks thin out, if not disappear altogether, over about the same time that we figure our planets formed. Even planet formation did not destroy all the solar disk, however. It still retains the Kuiper belt and immense amounts of interplanetary dust caused by comet disruptions and asteroid collisions. In the early 1980s, an orbiting infrared observatory found radiating disks of heated dust around several bright class A stars, including Vega and Fomalhaut. Several more have been found by Earth-based observatories, the best known Beta Pictoris, whose thick disk is seen edge-on. This stellar disk as well as others seem to be hollowed out, circumstantial evidence for planets.

Yet no extra-solar planet has ever been seen. Planets are relatively cold and shine at visual wavelengths only by reflection. Their feeble light is so overwhelmed by their stellar parents that they remain invisible. That limitation will change as observing capabilities by orbiting telescopes improve. Until that time, we must be content with a more indirect approach. Gravity is always mutual. As a planet orbits a star, so must the star orbit the planet, the ratio of movement inversely proportional to the ratio of masses. The planet moves a lot, the star hardly at all. But if the planet is sufficiently large and close, the stellar motion might be enough to be detected. The observation of actual positional shifts is difficult, and presently for most stars futile. Success was born instead from the detection of velocity—Doppler—shifts, which are terribly small, but measurable to about 3 meters per second, roughly the speed at which the Sun moves in response to Jupiter. Dozens of "planets" have been detected with this technique.

We humans, even astronomers, are nothing if not parochial. Not until 1543 did we understand that we were *not* at the center of the Universe, and that the Earth orbited the Sun. Not until our own time did we really understand that there was not one "big Galaxy"—ours—but that we are one of countless other galaxies, many bigger even than our own. We also thought that all planetary systems must be like ours. How wrong. The first known planet outside our own Solar System, one orbiting 51 Pegasi, set the stage for a puzzle yet to be entirely solved. With about a Jupiter mass, it orbits only 0.05 AU away from the star (12 percent Mercury's distance from the Sun) with a 5.5-day period. Several others are similar.

We have little idea how these planets are constructed. Perhaps they, or at least some, are brown dwarfs. As far as we know (and we do not know a lot), brown dwarfs are created whole from interstellar clouds, whereas planets are assembled from the bottom up, by the fusion of dust grains. The accepted lower limit to a brown dwarf is 13 Jupiter masses, below which not even deuterium fusion can be supported. Doppler analyses give only lower limits to orbiting bodies' masses, so some may fall into the brown dwarf range. There may also be considerable overlap, brown dwarfs made down to Jupiter-size or below, planets made up into the brown dwarf range.

If these bodies are truly planets and anything like Jupiter, made of hydrogen and helium, they seem to defy the theory presented for the formation of our planetary system, as they should never have been allowed to have formed in the high heat so close to their stars. Perhaps, however, they did *not* form there. Planets can move, change their orbits. Computer simulations show that as our giant planets tossed leftover planetesimals out of the planetary disk, they must have been gravitationally affected too, Jupiter moving somewhat inward, Uranus and Neptune moving rather significantly outward. A leftover thick dusty gaseous disk could have a similar effect, making a "Jupiter" spiral inward as a result of some kind of viscous friction. The planet would stop moving only after the disk dissipated. In some cases, a planet could spiral right into its star. Those we are picking up now are the lucky ones whose disks were removed just before disaster.

If nothing else, the discoveries are teaching us that a whole new science of extrasolar planets lies before us, and that planetary systems come in a great variety of forms. We have almost no idea of the details, and no idea of the extent. All we are doing now is finding the planets that nature allows us to see, the ones that are so large and so close to their parent stars, that they produce relatively large Doppler

Hubble captures thousands of stars within a dense portion of the Milky Way in Sagittarius. Each star has a different story to tell. Some are huge, some small, some young, some old. Many will play roles in begetting the next stellar generation, our own Sun and Earth a product of what has gone on before. At the beginning of the twentieth century we could only look and wonder. Now, we understand their meaning. (Hubble Heritage Team from data by J. Trauger, JPL; J. Holtzman, New Mexico State; et al.)

shifts. There may be many systems like ours out there, too, and many other kinds that we cannot even envisage. But someday we will find them; and more, someday we will actually see the planets themselves. Someday, too, we may actually know what is *on* them. Is anybody looking back at us wondering the same thing?

epilogue

From here, the Sun and other stars, some single, some multiple, some with planets, some without, will evolve. They will in their own time turn into giants or supergiants, some into planetary nebulae and white dwarfs, others into supernovae, neutron stars, black holes. In the process, they will add much of themselves, some further enriched in heavy elements of their own making, back into interstellar space. From this blend new stars will someday form, carrying with them the ashes of the Sun—and perhaps the ashes of planets too. Maybe we walk ground that was donated to the cosmos by an evaporated planet; perhaps someday someone on a planet orbiting another star will tread on ground that is partially made of Earth. So it has gone and will continue to go for unimaginable time.

We can link our Sun, our planet, ourselves, to the Universe, to the Big Bang, from which came the galaxies and their stars. We are made of stars, we belong to them and they to us. Look out to them as they sparkle overhead on a dark clear night. You *can* touch the stars. Just feel the good Earth beneath your feet.

index